FUNDAMENTALS OF SENSOR NETWORK PROGRAMMING

FUNDAMENTALS OF SENSOR NETWORK PROGRAMMING

Applications and Technology

S. Sitharama Iyengar
Nandan Parameshwaran
Vir V. Phoha
N. Balakrishnan
Chuka D. Okoye

A John Wiley & Sons, Inc., Publication

Published by John Wiley & Sons, Inc., Hoboken, New Jersey.
Published simultaneously in Canada.

Library of Congress Cataloging-in-Publication Data is available.

ISBN 978-0470-87614-5

Printed in Singapore

10 9 8 7 6 5 4 3 2 1

This book is dedicated to Professor Donald E. Knuth (Professor Emeritus at Stanford University) for his fundamental contributions to the programming in Computer Science.

Professors Daniel Siewiorek (Carnegie Mellon), John Hopcroft (Cornell University), Juris Hartmanis (Cornell University), Thomas Kailath (Stanford University), and K. Mani Chandy (Cal-Tech) have all inspired the authors for coming up with the first book on sensor programming.

Also, Dr. S.S. Iyengar would also like to dedicate this book to all his former and current Ph.D. students, his son Vijeth Iyengar and finally, grandson, Ranvir Iyengar. Professor Phoha would like to dedicate this book to Shiela, Rekha, Krishan, and Vivek.

—S.S. Iyengar
N. Parameshwaran
Vir Phoha
N. Balakrishanan
Chucka Okaye

Contents

Preface

The price of greatness is responsibility.
—Winston Churchill

Sensor processing is a central and an important problem in aerospace/defense, automation, medical imaging, and robotics, to name only a few areas. A surveillance system used in aerospace and defense is an example of a sensor processing system. It uses devices such as infrared sensors, microwave radars, and laser radars that are capable of detecting and tracking flying objects in their observational space. A sensor processing system may employ intelligent and disparate sensors that are distributed logically, spatially, and even geographically. It is then referred to as a *distributed sensor network* (DSN). The sensor may measure scalar values (e.g., temperature) or vector values (e.g., position in three-dimensional space). The measurements are generally a function of time and/or space. Because of variation in operating environments or other factors, such as aging and communication delays, the measurements may appear contradictory. Although combining the numerous sensor measurements may appear contradictory, it minimizes the uncertainty of measurements and improves reliability and fault tolerance.

There is a wide body of literature on sensor networks and on design, analysis, protocols, and other research-related issues in sensor networks. However, the issues of software development, in particular pedagogical material related to software development in sensor networks, has been left mostly untouched. We present this book to focus on software development in sensor networks. This book provides the basics needed to develop sensor network software and supplements it with many case studies covering network applications. We also examine how to develop onboard applications on individual sensors, how to interconnect these sensors, and how to form networks of sensors, although the major aim of this book is to provide foundational principles of developing sensor networking software and to critically examine sensor network applications.

Courses in sensor networks do not provide direct access to sensor networking equipment and software. However, the need for hands-on experience is essential for a high-quality education. We have structured the examples in the book, in such a way that a teacher can demonstrate the material on a small network of four (or more) sensor nodes and a laptop, all of which can be hand-carried to any room and set up on a teacher's desk. It is our hope to impart to students and by extension any reader of this book an understanding of sensor network programming based on general principles with an aim to develop actual software. Thus, in Part II we provide implementations of various algorithms ranging from very simple to more elaborate in complexity.

Features

This is a practical book. The book contains many figures, pseudocode, and actual code tested in a laboratory environment to explain the concepts. Most of the code is written and tested by N. Paramesh and our student Chuka Okoye, and some examples have been adopted and modified from nesC and TinyOS manuals. The programs were tested on various network configurations consisting of six TelosB sensor motes and a laptop. Currently, there exist several platforms designed to run programs written for TinyOS such as Mica, Telos, and Intel X-scale family of motes. Each of these motes has a unique functionality that differentiates it from others. While several of the programs that have been written to illustrate concepts in this book are platform-independent, most have been tested primarily on TelosB, a Telos family suite of sensors. TelosB motes are IEEE 802.15.4–compliant devices having an integrated onboard antenna and a range of sensing devices (light, temperature, and humidity sensors). They are characterized by their USB (universal serial bus) programmability; hence programs can be written on a computer and transferred to the device for testing. TelosB sensors have an already implemented 802.15.4 physical layer and a software MAC (media access control) layer that ensures that sensors can communicate with each other using the onboard antennas and existing radio drivers. Programs written in the TinyOS environment can be checked for program correctness either by running them in the TinyOS simulator (TOSSIM) or by using the integrated LEDs (light-emitting diodes) on the sensors for debugging. TOSSIM allows detailed simulation of TinyOS network stacks at the bit level, therefore allowing both low- and high-level applications that require fine-grained controls to be simulated. TinyViz, a graphical user interface (GUI) for TOSSIM, can be used to visualize and interact with running TOSSIM simulations. The TOSSIM simulator can mimic thousands of nodes and also has the ability to print debug information such as variables.

We make no pretense of providing in this book a reference manual for nesC or of TinyOS, and students writing more than trivial programs will do well to avail themselves of specific reference sources on nesC and TinyOS.

Audience

The book is suitable for upper-division (junior or senior) undergraduate-level or first-year graduate-level students. We have presented the material in this book assuming that the reader is knwoledegeable in some conventional programming language and has basic familiarity with an operating system. Although the programming language used in examples is nesC running under the TinyOS operating system, in some cases, particularly in Part III of the book, further knowledge of sensors and networks will be helpful. We have tried to arrange the material in an order that will facilitate following the chapters linearly. However, readers who are confident in the programming, data structures, have a basic understanding of sensors, and want to focus more on software development may skip to Chapter 7, Sensor Programming. In conjunction with the material in the book, readers are encouraged to work on moderate-sized projects taken from the book.

S. S. IYENGAR
Louisiana State University, Baton Rouge (USA)

N. PARAMESHWARAN
University of New South Wales (Sydney, Australia)

V. V. PHOHA
Louisiana Tech University, Ruston (USA)

N. BALAKRISHNAN
Indian Institute of Science, Bangalore (India)

C. D. OKOYE
Louisiana Tech University, Ruston (USA)

January 28, 2010

Foreword

Dr. Iyengar et al. have written an excellent piece of senior undergraduate or first year graduate level text for sensor or sensor network programming. The book contains numerous practical examples of real or pseudocode and will be extremely beneficial for both students and teachers. It will also be useful for working engineers writing code for sensor based complex or simple systems. Lately the use of sensors to measure space or temperature has grown exponentially from freeways to airports to medical devices to smartphones. Dr. Iyengar et al.'s book will have a positive impact to the industry in general.

Arup Gupta
Director Wireless Platform Technologies
Ultra Mobile Group
Intel Corp

Acknowledgments

This book started out with a series of discussions in 2008, and evolved from various research projects on sensor networks. It was contributed to and improved on by many individuals, and was funded by several federal agencies. The culmination of this work done by the LSU sensor network research group and others is recorded here. Sensor programming is very different from traditional programming. The authors have made an attempt to present the programming structure and implementation aspects of sensors under the framework of the nesC language.

Many researchers, friends, students, and faculty contributed to this work. We would like to acknowledge their contributions. This work has been supported in part by the DoD DEPSCoR grant by the Office of Naval Research and by a PKSFI grant, LEQSF (2007-12)-ENH-PKSFI-PRS-03, from the Louisiana Board of Regents. The authors would also like to thank the Indian Institute of Science for their travel support to the 100th Centenary Conference. Countless people have provided assistance; we have benefited from discussions with many of our colleagues around the country while working on this project. We are grateful to them all and sorry that we cannot list all by name, but a few need to be mentioned. We would like to thank V. Iyer for his several vital contributions to this work, including working with LaTex and general proofreading.We would like to thank Professor Ramamurthy and Professor M. B. Srinivas of the International Institute of Information Technology, Hyderabad, India for being co-authors of many previous publications that helped us put together this book for FARM (fusionable ambient renewable access control) applications.

We would like to thank the graduate students involved in proofreading/editing, including Robert, Rajesh, Srivathsan Srinivasagopalan, and many others. The authors would like to hear from any readers with comments, suggestions, or bug reports. In writing a book of this proportion, it goes without saying that many people have contributed both directly and indirectly to the inception, development, and completion of this project.

This endeavour would not have been achieved if not for the many hours toiled by both my present and former students. For this, thanks goes to Professors N.S.V. Rao (Oak-Ridge National Lab), Richard Brooks (Clemson), Mengxia Zhu (SIU-Carbondale), and Qishi Wu (U. Memphis) and current PhD students Robert Dibiano, Srivathsan Srinivasagopalan, Vasanth Iyer, J. Kim. Further, I want to thank the faculty at KAIST – S. Korea, specifically Dr. Song and his collaborators who participated in the early discussions of this project.

I want to thank all of my research collaborators, specifically, Krishnendu Chakrabarty (Duke), Sartaj Sahni (Florida), Bhaskar Krishnamachari (Southern Cal), and at Louisiana State University (Professors S. Mukhopadhyay, R. Kannan, C. Busch, J. Zhang, H. Wu, X. Li). Moreover, We, (Iyengar and Phoha) would like to thank various funding agencies, specifically DARPA, ONR, NSF, Army Research Office, Air Force (AFRL) and the Indian Institute of Science – Bangalore. Dr. Parameshwaran acknowledges the support and assistance needed from University of New South Wales, Sydney, Australia during his sabbatical at LSU. The authors also acknowledge Prof. Balakrishnan's research group at Indian Institure of Science, Bangalore, during the preparation of this manuscript. Prof. Phoha acknowledges the support of LaTech researchers and many of his graduate students: Kiran Balagani, Chuka Okoye, Sunil Babu, and colleagues Enam Karim and Rastko Selmic from Louisiana Tech University.

We would also like to thank Prof. Holger Karl and Dr. Adreas Willig, the authors of the book *Protocols and Architectures for Wireless Sensor Networks* (John Wiley & Sons, 2005) which was instrumental in providing the intial impetus for writing a book at a level with more programming details using nesC upon TinyOS.

Perhaps, most importantly, I would like to thank Dean Kevin Carman, for his continued support and mentorship in my research activities at Louisiana State University.

About the Authors

Each of the five authors has teaching and research interests in sensor networks. A brief biographies of these five authors follow.

Dr. S. S. Iyengar

S. S. Iyengar is currently the Roy Paul Daniels Professor and Chairman of the Computer Science Department at Louisiana State University. He heads the Wireless Sensor Networks Laboratory and the Robotics Research Laboratory at LSU. He has been involved with research in high-performance algorithms, data structures, sensor fusion, data mining, and intelligent systems. Since receiving his Ph.D. degree in 1974 from Mississippi State University (MSU), USA. He has directed over 40 Ph.D. students and 100 master's students, many of whom are faculty at major universities worldwide or scientists or engineers at national laboratories or industries around the world. He has published more than 380 research papers and has authored or coauthored six books and edited seven books, published by John Wiley&Sons, CRC Press, Prentice-Hall, Springer Verlag, and IEEE Computer Society Press, and other publishers. One of his books, titled *Introduction to Parallel Algorithms*, has been translated into Chinese. His research has been funded by the National Science Foundation (NSF), Defense Advanced Research Projects Agency (DARPA), Multi-University Research Initiative (MURI Program), Office of Naval Research (ONR), Department of Energy/Oak Ridge National Laboratory (DOE/ORNL), Naval Research Laboratory (NRL), National Aeronautics and Space Administration (NASA), US Army Research Office (URO), and various state agencies and companies. He has served on the US National Science Foundation and National Institute of Health panels to review proposals in various aspects of computational science and has been involved as an the external evaluator (ABET-accreditation) for several computer science and engineering departments. He is a fellow of (1) the Institute of Electrical and Electronics Engineers (IEEE), (2) Association for Computing Machinary (ACM), (3) American Association for the Advancement of Science (AAAS), and (4) Society of Design and Process Science (SDPS) and a member of the European Academy of Sciences. He has won many best-paper awards and IEEE Computer Society awards.

Dr. N. Parameshwaran

Nandan Parameshwaran is a senior lecturer in the School of Computer Science and Engineering, University of New South Wales, Sydney, Australia. He obtained his Ph.D. from the Indian Institute of Science, Bangalore, India. His research interests include multiagent system models, ontology-driven techniques for concept mappings, robust rule-based systems, and network management. He has been applying agent-based techniques for real-world applications, including social systems such as road traffic simulations, and fireworld scenarios. He teaches courses in Internet programming, functional programming, data structures and algorithms, and artificial intelligence. Dr. Parameshwaran has served as a member in the Florida Artificial Intelligence Research Society (FLAIRS) and the International Conference on Semantic Computing (ICSC) programming committees, and is a former member of both the IEEE Computer Society and the American Association for Artificial Intelligence (AAAI).

Dr. Vir V. Phoha

Vir V. Phoha is a professor of computer science in the College of Engineering and Science at Louisiana Tech University (USA). He holds the W. W. Chew Endowed Professorship at Louisiana Tech and directs the Center for Secure Cyberspace. He has won various distinctions, including ACM Distinguished Scientist, 2008; research commemoration awards at Louisiana Tech University (2002, 2006, 2007, 2008); outstanding research faculty and faculty Circle of Excellence Award at Northeastern State University, Oklahoma, and as a student was awarded the President's Gold medal for Academic Distinction. Professor Phoha holds an M.S. and a Ph.D. in Computer Science from Texas Tech University.

He has done fundamental and applied work in anomaly detection in network systems, in particular in the detection of rare events. His current research interests include autonomy and security issues in sensor networks. He has eight patent applications and many reports of inventions. He is author of over 90 publications and author/editor of three books: *Internet Security Dictionary*, Springer-Verlag (2002); *Foundations of Wavelet Networks and Applications*, CRC Press/Chapman Hall (2002); *Quantitative Measure for Discrete Event Supervisory Control*, Springer (2005).

Dr. N. Balakrishnan

N. Balakrishnan is a scientist of high international repute and is well decorated with prestigious awards. He received his B.E. (Hons.) in Electronics and Communication from the University of Madras in 1972 and Ph.D. from the Indian Institute of Science in 1979. He then joined the Department of Aerospace Engineering as an assistant professor. He is currently the associate director of the Indian Institute of Science and a Professor at the Department of Aerospace Engineering and at the Supercomputer Education and Research Centre. He played a crucial role in building India's first Supercomputer Centre and the National Centre for Science Information at the Indian Institute of Science.

His areas of research, in which he has published over 200 papers in international journals and for presentation at international conferences, include numerical electromagnetics, high-performance computing and networks, polarimetric radars and aerospace electronic systems, information security, digital libraries, and speech processing. He has received many awards, including the Padmashree, Homi J. Bhabha Award, the JC Bose National Fellowship, the Alumni Award for Excellence in Research for Science & Engineering at the Institute, the Millennium Medal of the Indian National Science Congress in 2000, Ph.D. (Honoris Causa) from Punjab Technical University in 2003, and the CDAC-ACS Foundation Lecture Award. He was the NRC Senior Resident Research Associate at the National Severe Storms Laboratory, Norman, Oklahoma (USA) from 1987 to 1989. He was a visiting research scientist at the University of Oklahoma in 1990, Colorado State University in 1991 and has been a visiting professor at Carnegie Mellon University since 2000. He is an honorary professor in the Jawaharlal Nehru Centre for Advanced Scientific Research (JNCASR).

He is a fellow of the Academy of Sciences for the Developing World (TWAS), Indian National Science Academy (currently the vice president), Indian Academy of Sciences, Indian National Academy of Engineering, National Academy of Sciences, and Institution of Electronics & Telecommunication Engineers. He is currently a member of the National Security Advisory Board, part-time member of the Telecom Regulatory Authority of India, and member of the Board of Governors of IIT Chennai. He is a directors of (1) Bharat Electronics Limited (BEL), (2) Data Security Council of India, and (3) CDOT-Alcatel Research Centre at Chennai; he is a member of the Council of CDAC and the council of many universities and CSIR laboratories. He is Editor of the *International Journal on Distributed Sensor Networks*. He was a member of the Scientific Advisory Committee to the Cabinet (SAC-C), a member of the Board of Governors, IIT Delhi; Chairman, All India Board of Information Technology Education of AICTE; and Editor of *Electromagnetics* and *International Journal of Computational Science and Engineering* until recently.

Chuka D. Okoye

Chuka D. Okoye is an undergraduate student majoring in computer science in his senior year at Louisiana Tech University. He has been named to the President's and Dean's lists multiple times for academic distinction and most recently nominated as the most outstanding computer science senior. During his 4 years at Louisiana Tech, he held various leadership positions, including the most coveted vice president of the local ACM chapter, president of the Robotics Club, and Louisiana Tech Programming Team Czar. His research interests include high-performance computing (HPC), machine learning, and wireless sensor networks. Having done exemplary work in the design and implementation of high-availability tools for clusters, he was charged with the development of a high-availability solution for the renowned Louisiana Optical Network Initiative (LONI). After being accepted in the Google Summer of Code Program under the mentorship of researchers from Oak Ridge National Laboratory, he greatly enhanced the widely popular OSCAR software stack for HPC clusters. Currently, he works as an undergraduate student researcher for the Center for Secure Cyberspace under the supervision and mentorship of the well-accomplished Dr. Vir V. Phoha in the Sensors and Machine Learning Group.

Notations and Abbreviations

Clarity of presentation is critical to effectively communicate knowledge.
—S. S. Iyengar

ACRONYMS*

AG	acquaintance group
API	application program interface
ARQ	automatic repeat request
ATR	atomic target recognition
BC	block-cut (e.g., block-cut tree)
BFS/DFS	breadth-first/depth-first search
BHS	baggage-handling system
CAN	controller area network
CBS	checked-baggage screening
CHAMP	caching and multipath (routing)
DIFS	distributed interframe space
DSN	distributed sensor network
DVS	dynamic voltage scaling
EEPROM	electrically erasable read-only memory
ESM	electronic support measure
FARM	fusionable ambient renewable MAC
FDMA/TDMA	frequency-/time-division multiple access
FFD	full-function device
FIFO	first-in/first-out
FSM	finite-state machine
GEAR	geographic and energy-aware routing
GHT	geographic hashing table
GPSR	greedy perimeter stateless routing
GSN	Global Sensor Network
GUI	graphical user interface
HAL	hardware abstraction layer

*Proprietary organization abbreviations (IEEE, etc.) and very common acronyms (e.g., CPU, GPS, IR, PC, UV) omitted from this list.

HBA	hub-based architecture
HNG	hop neighborhood group
HPC	high-performance computing
HSR	hierarchical state routing
IFF	IR (infrared) identification–friend/foe (sensor)
INSPIRE	innovation in sensor programming implementation for real-time environment
IPC	interprocess (or intermediate-performance) communication
ISR	intelligence–surveillance–reconnaissance
LEACH	low-energy adaptive clustering hierarchy
LPL	low-power listening
MAC	media access layer
MAS	multiagent system
MCU	multipoint control unit
MEMS	microelectromechanical system
MOP	maximal outer planar (graph)
OMNeT	objective modular network testbed
PLC	product lifecycle or public limited company
PSG	publish/subscribe group
QoS	quality of service
RAM/ROM	random-access memory/read-only memory
RFD	reduced-function device
RFID	radiofrequency identification (tag; as in, e.g., airport security)
RSA	random structures and algorithms
RSS	reallly simple syndication
RTC	run to completion
RTOS	real-time OS (operating system)
RTS	ready (or request) to send
SCP	scheduled channel polling
SHIMMER	sensing health with intelligence, modularity, mobility, and experimental reusability
SIFS	short interframe space
SMS	security management system
SPEED	systems planning, engineering, and evaluation device
TCP/IP	Transmission Control Protocol/Internet Protocol
TOSSIM	TinyOS simulator
UML	Unified Modeling Language
USB	Universal Serial Bus
UTC	Universal Time, Coordinated
WASN	wireless ad hoc sensor network
WSN	wireless sensor network

PART I
Overview

1 Introduction

> The creation of genuinely new software has far more in common with developing a new theory of physics than it does with producing cars or watches on an assembly line.
>
> —T. Bollinger

Software that drives the operations of sensors and communication among sensors is basic to any meaningful application of sensor networks. The goal of this book is to provide an understanding of how this software functions; how it allows the sensors to gather information, process it, and interact with each other in networks; and how these networks interact with the physical world. One aim of this book is to provide fundamental information necessary to write efficient sensor network software. A second aim is to provide a balance between theory and applications, so that the subject matter is complete (self-contained).

Wireless sensor network (WSN) applications may consist of diverse sensors with varying capabilities. For example, sensors may range from an extremely constrained 8-bit "mote" to less resource-constrained 32-bit "microservers." These sensors may be organized in different network configurations, which use different communication and data dissemination protocols, most software development platforms consist of libraries that implement message-passing interprocess communication (IPC) primitives, tools to support simulation, emulation, and visualization of networked systems, and services that support networking, sensing, and time synchronization. Given all of this diversity, there is an underlying theme of software development and deployment that cuts across platforms.

1.1 SOME FOUNDATIONAL INFORMATION

This section provides some basic information necessary for understanding the sensors and sensor networks.

1.1.1 Sensors

Typically a sensor is composed of components that sense the environment, process the data, and communicate with other sensors/computers. A sensor responds to a physical stimulus, such as heat, light, sound, or pressure, and produces a measurable electrical

Fundamentals of Sensor Network Programming: Applications and Technology, By S. S. Iyengar, N. Parameshwaran, V. V. Phoha, N. Balakrishnan, and C. D. Okoye

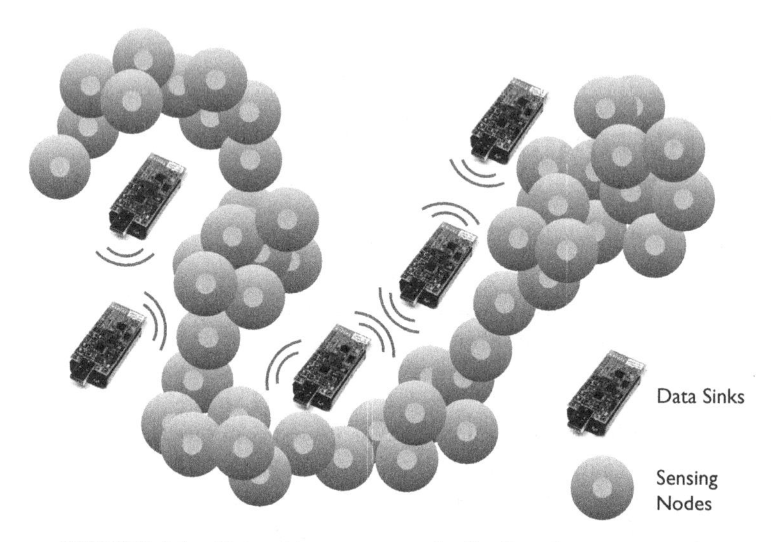

FIGURE 1.1 Networking structure of a distributed sensor network.

signal. Thus a sensor with its own sensing device, a memory, and a processor can typically be programmed with a high-level programming language, such as CorJava. The sensing devices can range from nanosensors to micro- and megasensors. In the remainder of this book when we refer to a *sensor*, we refer to a whole system such as a mote, which may have more than one physical sensor, its memory, processor, and other associated circuitry. Figure 1.1 shows a distributed sensor architecture and various components.

1.1.2 Sensor Networks

A *distributed sensor network* (DSN) is a collection of sensors distributed logically or geographically over an environment in order to collect data. Distributed computing and distributed problem solving are commonly used in DSN in order to abstract relevant information from the data gathered and derive appropriate inferences. This kind of data fusion can be used to compensate for the shortcomings of the individual sensor in real-world enviornments. For more details on sensor networks, see Refs. 1–3.

Most references to the term *sensor network* can denote multiple sensing configurations to be used in multiple contexts. Sensor networks typically consist of numerous sensing devices that may communicate over wired or wireless media, and may have as intrinsic properties limitations in computational capability, communication, or energy reserve. This does not imply that all sensor deployments consist of severely resource-constrained devices; for example, radar, closed-circuit cameras, and other wireline devices are commonly used in sensor network experimentations in academia and military research. These sensing devices possess reasonable computational capability and more importantly, may not have limited energy or constrained communication abilities. The main crux of this book is focused on the class of sensors having severely constrained computation, communication, and energy resources. These devices range from penny to matchbox in size and are deployed in an ad hoc and nonplanned (random) fashion. Examples of such devices include the mote platforms commonly used in academia.

1.2 NEXT-GENERATION SENSOR NETWORKED TINY DEVICES

1.2.1 Domain-Specific Challenges

Development of software in wireless sensor networks draws on experiences across several domains in computer science and some engineering disciplines such as

1. *Networking.* Networking knowledge is critical in sensor networks, providing information on how large-scale mobile ad hoc wireless networks can be created and managed efficiently.
2. *Power Systems.* Sensor networks, also rely on information from computer science and electrical and nanosystems engineering, in the creation of energy efficient software and hardware components, resulting in improved life of sensor networks.
3. *Data Management.* Experience in large-scale data management and data mining techniques is required in sensor networks since huge heterogenous datastreams are generated from these ubiquitous sensing devices.
4. *Data Fusion.* Since most devices have basic sensing capabilities, the need to create software systems capable of combining data from multiple sources to create more complex representation of the world is necessary; hence the need for data fusion. Fusion systems draw on advances in artificial intelligence, statistical analysis, and distributed systems.

1.2.2 Technology-Driven Methods

A few examples of technology driven methods in sensor networks follow.

1. *Flooding*, such as broadcast of packets in a synchronized network from source to destination until the path is formed to find the topology
2. *Clustering*, including K-means clustering to find K centers and form a cluster to minimize the distance between nodes in a dense region and efficiently form a topology
3. *Short-path algorithms* for data aggregation, such as data aggregation trees to form wireless spanners to efficiently collect data periodically
4. *Distributed algorithms* for energy and reusability loading and fault tolerance in large sensor networks

1.2.3 Wireless Sensor Network Environment

Sensor Network make it possible to monitor, instruct, or control various domains such as homes, buildings, warzones, cities, and forests. Sensor networks can observe the sensing environment at a close range and thus have many advantages, such as ability to monitor smallest details, proximity to places which are difficult to reach by humans, for example difficult terrain or hazardous environment. The major limitations of sensors are their limited power supply, limited communication bandwidth and range, and limited computation ability and memory capacity. Data transmission consumes a large percentage of energy; reducing the amount of data transmitted is the primary focus of data processing. The small bandwidth of the wireless links represents a challenge for data processing. Because of the limited communication radius of a sensor node, data may have to go through multiple hops to reach the final destination. This leads to extra power consumption in sensor nodes on the relay path. Limited processing and memory capacities restrict the complexity of data processing algorithms running at the sensor nodes. The intermediate results and other data are also burdensome to store in the node because of limited memory size. Sensor data are a stream: a real-time, continuous, ordered sequence with limited control over the order in which items arrive and the limitations of low battery life, low bandwidth, and low processing power and operating memory present programming challenges that are unique to the sensor network environment.

1.3 SENSOR NETWORK SOFTWARE

A network architecture and protocols are essential foundations for building software applications.

Developing computational/communication systems for deployment and application for wireless sensor networks has been a challenge since the mid-1990s. More Specifically, wireless ad hoc sensor networks have been largely designed with static and custom architectures for specific tasks, thus providing inflexible operation and interaction capabilities. WSN applications need to be programmed with constrained memory and process-centric resource requirements in mind, in order to write communication code with real-time sensing deadlines, which are critical to a dedicated scheduled measuring task. In short, the problem is the choice of abstraction for the sensor node runtime environment. Our computational framework or paradigm called INSPIRE, defines and supports nanofootprint and real-time deadlines, scheduled tasks for computing, and allows communication and sensing resources at the sensor nodes to be efficiently harnessed in high density event driven application-sensing fashion, through the use of an object oriented framework. A key feature of the runtime abstraction is that all the infrastructure used by the kernel is simulated to provide wireless communications using renewable energy resources with its unique extended lifetime model. This allows it to scale all the code to any processor. The implementation of INSPIRE on a target prototype node occupies less than 10–40 kB (kilobytes) of code memory; for details, refer to Chapter 10. The distributed source coding implementation is used to measure the sensing activity and memory overheads using traditional sensor applications without constraints, but more importantly, we highlight the reliability of the transmitted data from the measuring applications.

1.3.1 Technology-Driven Software

Individual updates of software are impractical because of the large number of nodes and the relative inaccessibility of deployed nodes. One solution for updating software in sensor nodes is the deployment of a support network of small, mobile, temporarily attachable nodes with virtual connections from a host PC to individual nodes. This scheme allows the use of standard tools to update the software in the individual sensor nodes. For many sensor networks in field applications, such as sensors deployed in unreachable places such as in water or trees, it is desirable to remotely update the software on the sensor nodes.

The following are a few issues to be considered when updating the nodes with software updates:

- Updates need to be planned. The items included in planning are tradeoffs of different updates relative to energy costs, the injection strategy for network configuration, and size reduction techniques that result in quick updates.
- Injection strategies of software. The strategies could include updating individual nodes, or sending updates to a base station or to a number of select nodes that may then disseminate the updates to other nodes.
- How software would be activated. Software may be either automatically activated or based on a set of rules, or manual activation may be required. To meet the requirements for backward/forward version compatibility, control over the order of node activation may be needed.
- Checking the downloaded software for integrity, version mismatch, and platform mismatch, and dynamically checking the operation of the downloaded software after it has been activated.
- Monitoring of update-related faults.
- Security-related issues, such as key distribution, authentication, secrecy, integrity, and authorization.
- Problems related to very small nodes, such as limited code memory, and almost no RAM or EEPROM (random-access or electrically erasable programmable read-only memory) for storing new code. Techniques may need to be developed for incremental building of new code into code memory (usually flash-RAM).
- Version control, that is, prevention version mismatch.
- Heterogeneity of sensor nodes. There can be various forms of heterogeneity; for example, there may be a mix of platforms, or a network may consist of a small number of "spine"/data backbone (shown to be optimal for data delivery) and a large number of lower-power nodes (for data collection). Here the backbone

nodes will have to handle different versions of their code base as well as different codebases.

- Performance. The time required to update nodes as well as tradeoffs between time and energy need to be considered.
- Provisions to recover from faulty updates, with mechanisms to verify the new software both before and during execution.

1.4 PERFORMANCE-DRIVEN NETWORK SOFTWARE PROGRAMMING

There are four basic issues here:

1. *Quality of Service.* In sensor networks quality of service is an important metric to analyze the performance and reliability of different WSN routing algorithms. As the sensor nodes use fixed batteries to sense and communicate, it is necessary to collaboratively use the network resources to minimize power usage and when idling, conserve power by using ultra-low-duty cycling. The communication module of a sensor mote uses a software "stack" and a radio to receive and transmit information. The network stack has many layers, spanning from physical layer to network layer; with various functionalities. By design, a running stack needs to use a small footprint and be power-aware, avoiding unnecessary overheads at every layer. The QoS can be defined as how the stack performs load balancing (reusability index), power-aware sleep scheduling (due to network density), and the reliability of sending sensed data wirelessly (at the datalink layer).
2. *Reusability Index.* This performance-based index can described as the number of times that a given node has been used as a clusterhead to communicate to a base station or a sink during its lifetime. As many of the clusterhead selection algorithms are distributed in nature, they will not overuse a specific node more than the critical number of times. If all nodes are used evenly, then the reliability of the network increases during the entire lifetime of the node.
3. *Sleep Scheduling.* Most of the deployed sensor network applications are dense because of the limited radio transmission range, so even when not transmitting, data nodes are subjected to overhearing and collision. These factors severely impact the total power consumed. So, in a dense deployment if a sufficient number of nodes are awake to receive the multihop traffic, then other nodes can shut off their radios after exchanging the next polling time, to minimize idling. By activating only a subset of nodes and scheduling timeslots for nodes to be active, sleep scheduling saves on power and avoids dropped packets. The end goal of each of these methods is to continue reciving data from the network for as long as possible.

4. *Datalink Reliability.* Data must be not only available but also accurate. In wireless sensor networks a node needs to not only communicate with its neighbors also forward the periodic sensed data over the network. Many of the MAC protocols are designed for efficient ad hoc communications but not for reliable data sensing as the radio does not have a way to filter floor noise or a new sensed value in harsh environments. For this reason, a twoway handshake is necessary between the MAC and the datalink layer, which allows them to reliably capture the new data everytime a data aggregation is performed. With this reliable datalink mechanism the clusterhead can further fuse the data from neighboring sensors and discard any false values.

1.4.1 Routing

In a sensor network stack the network layer is solely responsible for route planning and maintenance. Most of the energy used by the network is due to its routing activity. In implementing routing there are two methods, one at the network layer, which is controlled by distributed algorithms to form clusters and uses efficient clusterhead selection, and another at the MAC layer, which uses multihop routing to forward data at the lower layers by using best-effort QoS.

1.4.2 Data Aggregation

In a large sensor network deployment many parameters are sensed over a wide area and are periodically sent to the central coordinator. As the sensed parameters are the same at every node (similar sensor types are attached), WSN data aggregation allows reduction of the redundancy in a transmission by statistically evaluating the frequency of occurring samples and the trend direction that they have during its lifetime. When sensor nodes sample individually, only the aggregated data are transmitted, thus increasing the local processing and decreasing the radio usage per aggregation cycle. A simple example is using data compression at the nodes to send fewer bits during each transmission.

1.4.3 Security

Security is a constant threat to outdoor wireless environments; thus it is prudent to have an encryption algorithm that allows encryption and decryption of wireless communications. One novel way to implement a security algorithm is to have a oneway function which is NP-complete at the predeployment stage and cannot be decrypted with limited resources in a deployed site of operation. This method is more suitable in other applications of networks; in the case of WSN networks, because of the nature of their distribution one can design a network polynomial key that is not a local function. The broadcast message cannot be decrypted when a few nodes are compromised as it needs to have other parameters that are well distributed and concealed from the intruder.

1.5 UNIQUE CHARACTERISTICS OF PROGRAMMING ENVIRONMENTS FOR SENSOR NETWORKS

Sensor networks differ from both wired and wireless computer networks in many ways. The topology of sensor networks can change rapidly and frequently. The nodes in a sensor network do not have a global identifier such as an IP (Internet Protocol) address, and the number of sensor nodes in a sensor network may be an order of magnitude greater than that in a typical computer network. The memory and the processing capabilities of sensor nodes are limited in comparison to nodes in a computer network. These characteristics lead to a programming environment that is unique. Thus, the programs need to be short and efficient, providing capabilities of interfaces and links of components and modules to each other. Additionally, to save battery power, the nodes may need to have aggressive power management capabilities; thus the programming environment needs to provide mechanisms such as split phase, the nonblocking equivalent of common power-saving techniques such as the sleep command.

1.6 GOALS OF THE BOOK

The goals of this book are to develop programming methodologies unique to sensor networks, and present in an organized fashion techniques for programming of sensors to enable them to work effectively as a group. Thus, although the focus is on programming of the individual sensor, the goal is to enable the sensor to work within a collaborative environment.

1.7 WHY TinyOS AND NesC

TinyOS is an emerging platform that provides a framework for the most common type of sensor application programming. Thus we have a tool that can be implemented on small Crossbow sensors and a wireless sensor network that can be ported to different classrooms and laboratories. NesC provides a C-type, component-based language.

In NesC a module is the lowest level of component abstraction that implements any commands provided in its interface. It may directly address a particular hardware component such as a light sensor, providing methods that abstract the actual operation of that particular hardware component. Several modules may be grouped together using a configuration to form a larger component.

1.8 ORGANIZATION OF THE BOOK

The book is organized as follows. In Part I, we present an overview of the subject of sensor network programming, beginning with a general introduction in the remainder of this chapter (Chapter 1). Chapter 2 gives a general description of the wireless sensors. It explains the basic components of a sensor, its sensing environment, and the various roles that a sensor can play in a wireless sensor network. Chapter 3 discusses current sensor technology, including the major families and types of sensors currently in use, including the Mica, Telos, Tmote Sky families, and others.

Part II provides a general background for sensor network (SN) programming, beginning with discussions on data structures for sensor computing programming in Chapter 4. Sensor computing programming of individual sensors in a SN environment requires an understanding of data structures, such as arrays, queues, stacks, and lists, which are essential to programming. For network implementation, an understanding of graphs is useful to appreciate routing and message passing. Thus, Chapter 4 explains those data structures, which are essential for programming in a wireless sensor network environment. Chapter 5 explains the tiny operating system (TinyOS) environment, and is essential to understanding the subsequent chapters. It presents the structure of application programming interfaces (APIs) built using a nesC like structure, which facilitates the readability of the examples given in the rest of that chapter. For the sake of completeness and continuity, Chapter 5 also includes a bare-minimum description of nesC programming language. In Chapter 6, on nesC programming, the nesC language is formally introduced and some major concepts in the language are discussed.

Part III discusses and presents examples of sensor network implementation. Chapter 7 provides a basic introduction to sensor programming. It discusses some of the challenges encountered when programming large numbers of sensors and some interfaces provided by TinyOS to alleviate these programming challenges. Chapter 8, on algorithms for wireless sensor networks, is the core and the major focus of the book. It gives detailed descriptions of various algorithms and their implementation in nesC. In Chapter 9, on techniques for protocol programming, we discuss several protocols used in most wireless sensor networks and provide accompanying pseudocode to explain the concepts.

Part IV presents real-world scenarios in sensor network programming. In Chapter 10 we discuss some programming abstractions that simplify the development and deployment of sensors. Chapter 11 presents standards for building WSN applications, with a brief overview of the ZigBee networking standard. Chapter 12 discusses an active sensor approach to distributed algorithms, widely known as INSPIRE (innovation in sensor programming implementation for real-time environments). Chapter 13 explores the performance analysis of networks in some detail with respect to power-aware algorithms. Chapter 14 describes sensor network modeling through design and simulation. This chapter presents an architecture of a sensor simulator and a sensor node that is used in the simulator, and further elaborates that OMNeT++ is a viable discrete-event simulation framework for studying both the networking aspects and the distributed computing aspects of sensor networks. We present the architecture of a sensor node that is used in the simulator and the general architecture of the simulator. Chapter 15 presents a MATLAB implementation of simple data processing and decisionmaking logic to be used to detect and respond to events in an airport baggage-handling system. Chapter 16 consists of closing comments.

1.9 FUTURE DEMANDS ON SENSOR-BASED SOFTWARE

In the future, advances in microelectromechanical systems (MEMSs) will lead to miniature sensing devices of about 20 (μm micrometers) to a millimeter in length. These devices will be self-powered, allowing even more collaboration with other devices. In regard to software, more standards specifying how data can be exported between different sensor networks will be established, allowing a more enriched and integrated sensing experience such as

- Real-time collaboration between navigation systems and traffic monitoring sensors
- Current information about seat availability at local restaurants or physicians offices
- Real-time environmental awareness by a wide range of applications and devices leading to better management of scarce resources, such as smart energy-saving homes.

In this regard, the principles addressed in this book will serve as building blocks for developing large-scale, longlived systems requiring self-organization and adaptivity.

PROBLEMS

1.1 Define the following:
  (a) Sensor
  (b) Ad hoc network
  (c) Distributed sensor network
  (d) Wireless sensor network
  (e) Reusability index

1.2 Discuss some of the design challenges that set wireless sensor networks apart from conventional networks.

1.3 Crossbow Technology Inc.'s MTS400 multisensor board is one of the most popular multipurpose heterogeneous sensing devices available on the market. Research and prepare a two-page report discussing the specifications and functionality of the MTS400 multisensor board.

1.4 Write a one-page summary of the article by Akyildiz et al. [4].

1.5 Other than those discussed in this introductory chapter, list three advantages and three disadvantages of sensor networks.

1.6 Crossbow Technology Inc.'s MICAz mote and Europe's Smart-Its platform are two popular sensor platforms. Research and contrast the features of MICAz with smart-Its (in terms of size, weight, battery life, onboard sensors, memory, CPU, operating system, processing limits, radio range, etc.).

1.7 What are the unique characteristics of programming environments for sensor network software?

1.8 Other than sleep scheduling, give two techniques to conserve the battery life (energy) of nodes in a sensor network.

1.9 State the issues to be considered when updating computation/communication software in a sensor network.

1.10 Does the data aggregation strategy adopted by a sensor network application affect its operational integrity and security? If "Yes," explain how and if "No," explain why.

1.11 In about five paragraphs discuss any three of your favorite real-world sensor network applications.

1.12 Sensors mounted on moving objects can open many interesting real-world applications. For example, sensors are already mounted on devices such as mobile phones to sense temperature, motion, and other parameters. Suggest some applications where mounting sensors on mobile objects will be useful.

1.13 Interesting scenarios are created when sensors are made much smaller in size and are programmed to become more autonomous. "Smart dust" refers to tiny devices that are capable of limited sensing, computations, and commmications capabilities, with short lifetime. Suggest some applications where one or many "bags" of smart dust can be used.

1.14 When wireless sensors become tiny and are deployed in very large number (such as in several bags of smart dust), interestingly, the overall behavior of such a system will in some way behave like social systems, exhibiting autonomy, self control, limited lifetime, and intracommunications. Identify the management challenges in such a social system. Consider an example application, and propose specific management solutions appropriate for this application.

1.15 Traditional programming views programs as a mapping from input values to output values. Suggest some characteristics of programs written for the wireless devices. [*Hint:* A sensor program spends a considerable amount of time in communication, and thus must be sufficiently ingenious to manage its resources (such as data and power), and cooperate with other sensors to achieve the overall behavior as required by the application.]

1.16 In a computer network, each node communicates with the other nodes using a set of protocols. What will be the limitations of this model when applied to wireless sensor networks? Suppose that we augment the protocols with flexible dialog features where each sensor node engages "intelligently" with the other sensor nodes. What will be the advantages? Discuss the resulting overhead.

1.17 Suppose that we view a WSN as a multiagent system (MAS). Suggest an application where this view will be appropriate. What will be the undesirable aspects inherent in such a MAS model?

1.18 Traditional network systems are designed to satisfy strict specifications. Because of the dynamic and uncertain environments in which WSN is employed, traditional approaches may not be appropriate. Investigate why this may be the case. If self-autonomy is one possible solution, how can it help the sensor network in satisfying the application requierements? What additional complications will this solution create for the application?

1.19 Indentify some aspects of security issues that are unique to WSN but may not be present in the traditional computer network systems.

REFERENCES

1. R. R. Brooks and S .S. Iyengar, *Multi-Sensor Fusion,* Prentice-Hall, Englewood Cliffs, NJ 1997.
2. K. Chakrabarty and S. S. Iyengar, *Scalable Infrastructure for Distributed Sensor Networks,* Springer-Verlag, 2005.
3. S. S. Iyengar and R. R. Brooks, eds., *Distributed Sensor Networks,* CRC Press, Dec. 2004.
4. I. F. Akyildiz, W. Su, Y. Sankarasubramaniam, and E. Cayirci, A survey of sensor networks, *IEEE Commun. Mag.* **40**(8):102–114 (Aug. 2002).

2 Wireless Sensor Networks

The Eight Fallacies of Distributed Computing—"Essentially everyone, when they first build a distributed application, makes the following eight assumptions. All prove to be false in the long run and all cause big trouble and painful learning experiences."

1. The network is reliable
2. Latency is zero
3. Bandwidth is infinite
4. The network is secure
5. Topology doesn't change
6. There is one administrator
7. Transport cost is zero
8. The network is homogeneous

—Peter Deutsch

A *sensor* is a device that responds to a physical stimulus (heat, light, sound, pressure, etc.) and produces a corresponding measurable electrical signal. Sensor systems have been used in military, industrial, and medical applications for many years. In military applications, sensor systems are employed for tasks such as ocean surveillance and in air-to-air defense which detect, track and identify the targets and events. These defense systems use sensors such as radar, passive electronic support measures (ESMs), infrared identification-friend foe (IFF) sensors, and electrooptic image sensors. In nonmilitary areas sensor systems are widely used in applications such as robotics, automated control of industrial manufacturing systems, smart buildings, traffic control and management, monitoring organs of the human body, and surveillance of natural disasters.

More recent advances in micromechatronics systems and microfabrication technology have led to the availability of low-cost, low-power, multifunctional modern sensors. A modern sensor consists of a sensing device(s), micro-controller, onboard memory, and a transceiver. The structure of a sensor node is dependent on the application. In general, a sensor node consists of four basic components: sensor unit, processing unit, transceiver unit, and power unit. A sensing unit is usually composed of sensors and analog-to-digital converters (ADCs). The analog signals are produced

Fundamentals of Sensor Network Programming: Applications and Technology, By S. S. Iyengar, N. Parameshwaran, V. V. Phoha, N. Balakrishnan, and C. D. Okoye

by the sensors that observe the phenomenon. These signals are converted to digital signals by the ADC, and then sent into the processing unit. The processing unit instructs the sensor node to carry out the assigned sensing tasks and manages the sensor node's collaboration with the other nodes. A processing unit has enough storage for the real-time operating system, protocols, and other application-specific algorithms. The transceiver unit connects the node to the network. The power unit supplies the energy for the sensor node, which may be batteries (coincell, lithium, etc.) or solar cells. According to the design purpose, sensor nodes can be divided into three categories: augmented general-purpose computers, dedicated embedded sensor nodes, and system-on-chip nodes. A sensor node of the first type normally has a higher volume than do the other two types, and relatively higher processing and memory capabilities. Off-the-shelf operating systems such as Linux and Win CE are run in real time, and standard wireless communication protocols such as IEEE 802.11 or Bluetooth are used in the node. A wide range of sensors from simple microphones to sophisticated videocameras can be accommodated on the node. Examples of this type of nodes include Sensoria WINS sGate nodes and various personal digital assistant (PDA) devices.

Dedicated embedded sensor nodes have a low compact volume as one of the design objectives. They have limited processing and memory capacities. Thus, special operating systems such as TinyOS and companion programming languages have been developed. Sensor nodes of this type include the Berkeley mote family and SunSPOT. Because of the low-cost and low-volume features, the future widespread use of such nodes is expected.

A system-on-chip node appears to be the direction of future sensor nodes, where extremely low power and small size are achieved. Such types of nodes will enable new applications to wireless sensor networks. It is envisaged that, for example, the tiny nodes can be mixed in the paint material and painted on the surface of a bridge to monitor the safety of the bridge. The design of such nodes needs to find a new way to integrate MEMS, complementary metal oxide semiconductor (CMOS) and radiofrequency (RF) technologies. Researchers at University of California, Berkeley have developed a prototype of system-on-chip node, called Spec Mote. The mote combines ultra-low-power computation, communication, and sensing into a small [only 5 mm^2 (2×2.5 mm)] single chip.

Networked sensors broaden our capability to observe the physical world. Sensor networks enable us to observe objects at close range and provide the possibility of monitoring previously unobservable phenomena. Satellites and radars are large and powerful sensors that have been widely used. This type of sensor has long-range sensing ability, and a single sensor can detect or monitor a large area. However, such sensors are not suitable for an environment where the line-of-sight paths are very short, for example, detecting individual animals in a forest, or monitoring a patient in a hospital. Small sensors can be used in such situations. Since each sensor node has a short sensing–transmitting range and limited processing ability, a number of nodes are networked to accomplish complicated tasks. As the sensor nodes can be placed close to or in the objects, greater informational accuracy is achieved and some previously unknown phenomena may be discovered. Additionally, large numbers of networked sensors can improve the reliability of sensing tasks. It is possible to densely deploy sensors because of their small size and low cost. If a sensor node fails, other sensors close to it can accomplish its task by working collaboratively. This is especially useful in environments where it is impossible to replace sensors. Sensors working collaboratively can also benefit resource conservation in sensor networks.

Sensors can be networked through physical connections or wirelessly. In early sensor systems, few sensors were deployed closely, such as in industrial control or medical monitoring, and were connected through wires to a control unit. With low-volume, low-cost, and wireless communication, the sensors can be deployed in a large area. Even in densely deployed systems, wireless communication reduces the complexity of wiring. Thus, wireless sensor networks have emerged as a new information-gathering paradigm with large numbers of sensors collaborating. We next describe sensor network applications briefly, followed by the characteristics of sensor networks. As our focus is on data-processing issues, the nature of sensor data is examined in this section.

2.1 SENSOR NETWORK APPLICATIONS

Sensor networks are deployed for collecting information on entities of interest. The availability of low-cost, low-power, and multifunctional intelligent sensors enables military applications such as battlefield surveillance; nuclear, biological, and chemical attack detection; and civilian applications such as habitat monitoring, environment observation and forecasting, health applications, vehicle traffic management, and smart environments.

2.1.1 Sensors

Habitat monitoring is regarded as a driver application of wireless sensor networks, and benefits the scientific community and facilitates ecological protection. The use of sensor networks eliminates the potential impact of human presence, and enables data collection at scales and resolutions that are difficult to achieve through traditional instrumentation. A notable research project is monitoring seabirds on Great Duck Island (in Frenchboro, Maine), a project conducted by the researchers from the University of California, Intel Research (Fig. 2.1).

The UC Berkeley Mica motes are deployed to collect data for studying seabird behavior during breeding seasons, effects on the environment of nesting seabirds, and related phenomena. Research on tracking and controlling animals has also been conducted. Wireless sensor networks have also been introduced for monitoring plants. PODS is a research project at the University of Hawaii to test ecological environments with rare and endangered species. The puspose of this research is to determine why endangered species of plants grow in one place but not in neighboring areas. Instead of managing sensor networks in untraversed places remotely, researchers from Intel Research deployed sensors in a vineyard and explored the future usage of sensor networks in agriculture. They investigated sensor network design while considering

FIGURE 2.1 On Great Duck Island in Maine, wireless sensors at 1 and 2 pass data to a gateway node shown at 3. It is then passed to the base station at 4, and potentially sent over the Internet via the satellite dish at 5.

the structure of work activities in agricultural production environments. As a result, some activities can be completed much more efficiently. For example, a vineyard can be sprayed only in places where there is a risk of powdery mildew. Grape Networks has transferred such research achievements into commercial products. Using Internet and mobile wireless mesh sensor networks, farmers can monitor and receive alerts with a PC, PDA, or mobile phone on the environmental information such as soil moisture, microclimates, and diseases in vineyards or open fields.

2.1.2 Sensor Networks

Developments in geographical information system (GIS), satellite remote sensing, and global position satellite/system (GPS) technologies helps people receive disaster-alert information more rapidly and accurately. These techniques have disadvantages such as low resolution and weather interference because of long distances. With the use of small and low-cost sensors, a large number of sensors can be spread in situ. Various environmental data can be collected in real time, and monitoring resolution can be greatly improved. Research on the use of wireless sensor networks for forest fire detection has attracted considerable attention from the research community. Firebug is a small, wireless sensor for collecting real-time data in forested areas developed by UC Berkeley. Equipped with a position system device, this sensor can collect data such as relative humidity, temperature, and pressure in situ and communicate with the remote data server through base stations. Thus, users can use the Internet to monitor the corresponding area. The relative risk of forest fire danger with respect to sensor measurement data such as smoke, windspeed, and light level has been investigated. Flood detection is another application in environment observation. An alert system is used to evaluate the possibility of potential fooding. This system is equipped with wind, temperature, and water-level sensors, and is able to provide real-time water-level and rainfall information. Wireless sensor networks also can be used for atmosphere pollution monitoring and detection of other natural disasters such as earthquakes and tsunamis.

2.1.3 Health Applications

Hospital and medical facilities can be improved by smart sensors. For example, patients with sensors on their bodies can be monitored and tracked inside a hospital or be talemonitored in their homes. Medical applications have unique demands such as extreme robustness, very dense networks, and preserving the privacy of medical data; research has been conducted specically for this type of application. CodeBlue is a project at Harvard University to explore wireless sensor network technology in a range of medical applications, such as disaster response, prehospitalization and in-hospital emergency care, and stroke patient rehabilitation. In this project, a software infrastructure was developed to provide routing, discovery, and security for wireless medical sensors, PCs, PDAs, and other devices, so that patients can be monitored and treated in a wide range of medical settings.

2.1.4 Vehicle Management

Inexpensive wireless sensor networks enable many applications in vehicle management such as traffic control, vehicle tracking and detection, monitoring car theft, and parking management. The current vehicle traffic – monitoring systems (e.g., see Fig. 2.2) use some buried sensors, cameras, and an associated communication network, which are expensive and generally limited to a few critical points. In the future, cheap sensors with networking capability can be attached on/in vehicles and deployed at every road intersection. Thus, traffic information such as the speed and density of traffic and the location of traffic jams, may not only be obtained by the traffic control center but also exchanged among vehicles passing each other. Further, sensors installed in a vehicle and communicating with those in other vehicles can help reduce the likelihood of vehicle collision. Deploying sensors in the pavement can assist with traffic control of pedestrians and vehicles.

2.1.5 Smart Environments

Deploying sensor networks in indoor environments such as buildings, homes, offices, and laboratories can make the environment seem "alive." Such smart environments can track and record the activity of people in that area or actively react and interact with people. Examples of research conducted in this area include home automation, residential laboratory building environmental control, interactive museum, and smart kindergarten applications. Embedding smartsensors in structures such as bridges can monitor the usage status and infrastructure security.

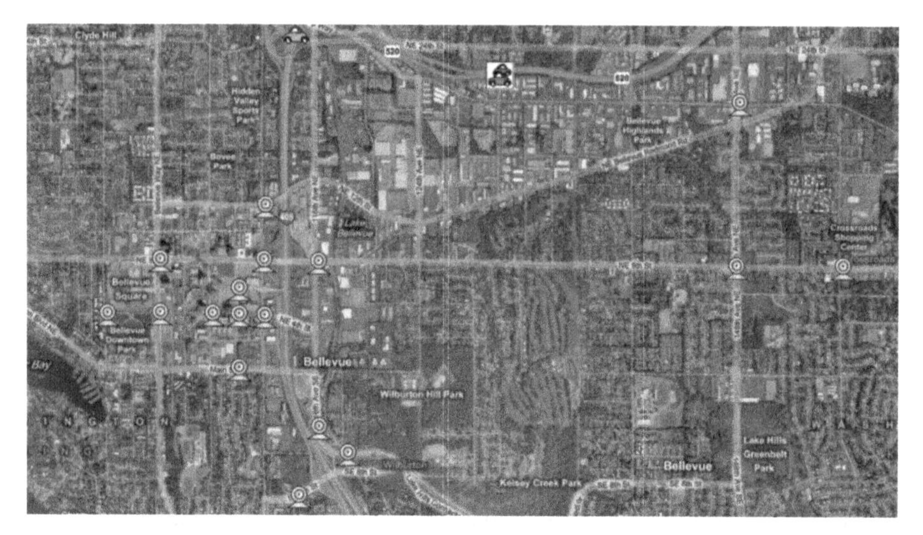

FIGURE 2.2 Traffic-monitoring sensor network on Microsoft SensorWeb.

2.2 CHARACTERISTICS OF SENSOR NETWORKS

Wireless sensor networks are capable of observing the enviornment, transferring data among their nodes, and making decisions on the basis of these observations. These networks are important for a number of applications—most notably for target detection and localization, surveillance, and enviornmental monitoring. Advances in miniaturization of microelectronic and mechanical structures have given rise to many battery-powered sensor nodes that have sensing, communication, and processing capabilities. These sensor nodes can be networked in an ad hoc structure to form distributed sensor networks for sensor processing in a distributed manner. Such networks have greater fault tolerance and sensing accuracy, and are less expensive than are traditional networks. Another interesting property of wireless sensor networks is that nodes can be deployed in hostile environments to provide continuous monitoring and processing capabilities for a broad variety of applications. More importantly, a sensor node integrates hardware and software for sensing, data processing, and communication. For more details on sensor integration, see an earlier text by the author [3]. Furthermore, these nodes can be deployed in large numbers in unstructured environments; the nodes in these networks rely on wireless channels for the transmission and receipt of data from other nodes. Communication among nodes can be characterized by the type of sensors that make them up; for example, RF sensors used in Berkeley motes have a maximum operation range of around 100 ft. One interesting property of wireless sensor networks is a parameter termed *sensing area*, which depends significantly on the physical types of sensors being used. For example, a *range sensor*, such as a range-polarized 6500 ultrasonic ranging module used in many robotics applications can detect a target from as close as 6 inches (in.)

away up to a maximum distance of 35 ft. Figure 2.2 shows an example of a distributed sensor network.

Wireless sensor networks are a subtype of ad hoc networks, which are well studied. However, from a design standpoint, wireless sensor networks are very unique compared to standard networks. The nodes are potentially mobile and/or unreliable, and we need to approach these networks from a totally fresh viewpoint in order to analyze them. Akyildiz et al. [1] highlighted some of the main differences between wireless and traditional ad hoc networks in a list similar to the following:

- The number of nodes in a WSN may be several orders of magnitude higher than in traditional networks.
- Sensor nodes are prone to failure.
- Sensor nodes are limited in energy, computational capacities, and memory. Consequently WSNs cannot afford table-driven MATNET protocols requiring too much memory to store routing tables.
- As opposed to the one-to-many broadcast model common in ad hoc networks, WSNs often use a many-to-one communication model with the topology of the reverse multicast tree.
- Sensor nodes are densely deployed.
- The topology of a WSN changes frequently.
- Sensor nodes may lack global ID.

Because of these differences, studies on WSN have developed into a seperate and distinct domain from conventional network development.

WSN Protocol Stack Akyildiz et al. presented a protocol stack for WSN [1], which they modeled after the ISO OSI (International Organization for Standardization open system interconnection) model. It has five layers: physical, datalink, network, transport, and application, and three *planes*: energy management, mobility management, and task management as shown in Fig. 2.3. In this model, a stack would be maintained across multiple nodes on the network, with sinks/clusterheads/elected leaders implementing the lions share of the stack.

The *physical layer* deals with A/D–D/A conversion, modulation/demodulation, and transciever techniques such as RF carrying. The datalink layer is concerned mainly with media access control and error recovery. The *network layer* deals with routing; because of the one-to-many and many-to-one data transmissions that are so common in a WSN, standard routing protocols are not very useful here. On top of this, network protocols seldom consider power usage and its effect on network lifetime, both critical factors in WSN design. Because of power and processing limitations, WSN cannot afford to implement standard routing tables and generally must adopt some lightweight alternative. The *transport layer* provides ports or transport interfaces to various applications just as in the ISO OSI model. Finally, the *application layer* decomposes and implements application-specific tasks by utilizing the lower layers.

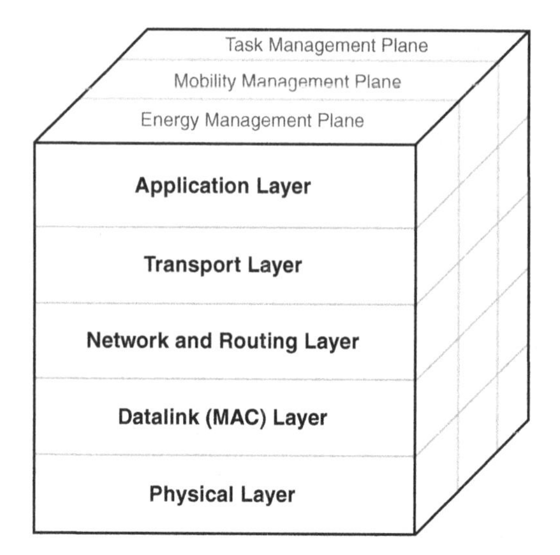

FIGURE 2.3 WSN protocol stack.

This protocol stack is just a simple model to allow us to analyze the problem of WSN design from a new paradigm; the proper relationship among the management planes and between management planes and network layers is undoubtedly more complicated. Nevertheless, this model is a useful design tool.

The energy management plane covers network wide energy conservation, and is reponsible for energy-saving decisions and tactics to increase network lifetime. The mobility management plane is responsible for keeping track of the physical location of network nodes, and responding to changes in this layout. The task management plane schedules and manages tasks. For more details on this topic, see the text by Chakrabarty and Iyengar [2].

Resource-Constrained Computing Environment Small size and low cost enable sensors to be deployed in a large number in various applications. Such features lead to the following resource constraints:

- *Energy.* The power supply of a sensor node is from batteries such as AA, coin cell, and lithium batteries. Thus the energy supply to sensor nodes is limited.
- *Communication.* sensor nodes communicate over wireless links with limited bandwidth, which is of the order of a few hundred kilobits per second (kbps).
- *Computation.* Sensor nodes operate with limited processing ability and memory capacity. For instance, the processor speed of a Mica mote is only 4 MHz and its random-access memory (RAM) is 4 kB.

Dynamic Topology Although a sensor network is usually deployed with stationary sensor nodes, the network topology is prone to frequent changes. This may be caused by the failure of nodes or links, power running out, or sometimes the mobility of nodes. New nodes may be added or old nodes removed from the sensor networks. As a result, techniques such as dynamic route changing are needed to adapt to network topology change.

Unpredictability *Unpredictability* in sensor networks refers to the uncertainty in the correctness (accuracy) of sensor data, the reliability of communication links, and the connectivity of networks. Sensor networks are subject to such uncertainty from many sources. The correctness of sensor data can be affected by the node status and transmission situations. Sensor nodes may fail because of lack of power or may be damaged by the uncontrollable events in the natural world, such as fire and earthquakes. Some types of sensors may need to be recalibrated after running for a certain period of time. Environmental interference can also cause unpredictable readings from sensors. The nature of sensor signal propagation and the environment for the signal propagation such as the surface roughness and the presence of reacting and obstructing objects influence data communication in wireless sensor networks. The wireless communication links shared by densely deployed or high-traffic-density nodes are subject to heavy congestion and jamming. High bit error ratio, low bandwidth, and asymmetric channels make communication unpredictable. Because of the continual possibilty of loss of nodes and links due to power constraints, damage, or eventual failure, the connectivity and routing structures of the network will change dynamially. This unpredictability represents many challenges for sensor networks, including the design of communication protocols, and development of data management techniques.

Heterogeneity Early research in wireless sensor networks focused on homogeneous architecture, in which all sensor nodes possess identical software and hardware. This architecture is resilient to individual failures. More recently, particularly in real-world deployments, heterogeneous sensor networks have become popular because of the advantage of increasing network lifetime and reliability. Heterogeneity in sensor networks can be grouped into three categories: *computational heterogeneity*, where nodes have different computational power; *link heterogeneity*, where some nodes have long-distance, highly reliable communication links; and *energy heterogeneity*, where some nodes may have unlimited energy resource such as being connected to a wall outlet. The network architecture of heterogeneous sensor networks likely has several tiers of nodes with different performance characteristics. Heterogenous networks pose challenges to data-processing techniques due to different data semantics and volumes, to communication protocols due to various links, and to security control due to different computational purposes of sensor nodes.

2.3 NATURE OF DATA IN SENSOR NETWORKS

The data produced by sensors in sensor networks have unique characteristics because of the goal of recording changes or rare events in many applications, the nature of sensor nodes, and the deployment of networks.

Streaming Sensor nodes produce data continuously. A temperature sensor may produce data every second, and can be represented as a tuple: < time; sensor ID; value >. This amounts to over 2 million tuples per month. For a large network of sensors, the total data will be in the order of terabytes per month, too much to store. This streaming nature raises several requirements for data processing. Sensor data must be processed online, and in-network processing is necessary. Further, sensor data should be appropriately precomputed and stored in a convenient format for later queries.

Correlation Spatiotemporal correlation exists among sensor observations. Because of the limited communication range of a sensor node, a wireless sensor network requires a spatially dense deployment of sensors to achieve satisfactory coverage. As a result, the sensor observations about a single event from multiple sensors are spatially proximal, and thus are spatially correlated.

Uncertainty The unpredictability of sensor nodes and links in sensor networks leads to sensor data uncertainty. Sensor data might not be delivered at reliable rates, the data may be incorrect as a result of sensor node damage, incomplete because of packet loss, or inaccurate owing to environmental interference. Techniques have been developed to derive proper information from the data obtained with uncertainty.

Heterogeneous Sources Data from heterogeneous information sources pose challenges to data processing techniques in wireless sensor networks. In practice, different sensors will be deployed to obtain information from different sources, so different data formats and semantics should be considered in data processing. Data from static data sources may be used along with sensor data for answering queries.

PROBLEMS

2.1 List and discuss some WSN applications

2.2 Briefly discuss the three types of sensor nodes.

2.3 Explain how you would set up a real wireless sensor network to monitor the temperatures in four corners of a room. List the equipment (sensor boards, base stations, interfaces, etc.) that you would use and detail the steps you would follow to set up the sensor network.

2.4 Write a one-page summary of the article by Szewczyk et al. [4].

2.5 Prepare a one-page summary of the article by Lorincz et al. [5].

2.6 Write a two-paragraph (~500-word) summary of the article by Essa [6].

2.7 Discuss the characteristics of sensor networks. Contrast them with the characteristics of computer networks (e.g., the Internet).

2.8 Choose your favorite real-world wireless sensor network application and list five causes/sources of its unpredictability.

2.9 Research and prepare a two-page report describing a dynamic routing strategy for a wireless sensor network. Discuss its strengths and weaknesses in detail.

2.10 Describe a real-world heterogeneous wireless sensor network application.

2.11 Discuss the data management issues in wireless sensor networks.

2.12 Consider the habitat-monitoring application discussed in the text (see also Fig 2.1). For this application, consider the scenario where the sensors are dropped from an aircraft over a large area on the ground periodically. An important aspect in such scenarios relates to the management of the sensors and the resulting network. Compare the resulting network characteristics with the standard computer network characteristics, particularly with respect to the following attributes (discussed in Section 2.2).

(a) The number of nodes in the network

(b) Node failure

(c) Communication across nodes

(d) Density of deployment

(e) Topology

(f) Node IDs

2.13 An interesting situation arises when sensors are mounted on tiny moving objects such as robots, obviously increasing the cost of the network. For the habitat-monitoring application mentioned above, comment on the network characteristics with respect to the following issues:

(a) Can you still drop the sensors from the aircraft?

(b) What can you do when a sensor is not able to communicate with its neighbor?

(c) How does it affect the density of deployment?

(d) What can you say about the topology of the network?

(e) How important is the node ID for a node in this situation?

REFERENCES

1. I. F. Akyildiz, W. Su, Y. Sankarasubramaniam, and E. Cayirci, Wireless sensor networks: A survey, *Comput. Networks* **38**:393–422 (2002).

2. K. Chakrabarty and S. S. Iyengar, Scalable Infrastructure for Distributed Sensor Networks, Springer-Verlag, 2005.

3. S. S. Iyengar, L. Prasad, and H. Min, Advances in Distributed Sensor Integration: Application and Theory, Prentice-Hall, Englewood Cliffs, NJ, 1995.
4. R. Szewczyk, E. Osterweil, J. Polastre, M. Hamilton, A. Mainwaring, and D. Estrin, Habitat monitoring with sensor networks, *Commun. ACM* **47**(6):3440 (June 2004).
5. K. Lorincz, D. J. Malan, T. R. F. Fulford-Jones, A. Nawoj, A. Clavel, V. Shnayder, Sensor networks for emergency response: Challenges and opportunities, G. Mainland, M. Welsh, and S. Moulton, *IEEE Pervasive Comput.* **3**(4):1623 (Oct.–Dec. 2004).
6. I. A. Essa, Ubiquitous sensing for smart and aware environments, *IEEE Personal Commun.* **7**(5): 4749 (Oct. 2000).

3 Sensor Technology

> Never before in history has innovation offered promise of so much to so many in so short a time.
>
> —Bill Gates

A typical wireless sensor network consists of spatially distributed sensors that co-operatively monitor some physical phenomena. This network formed by the sensors could contain nodes with varying capabilities and sometimes completely different underlying platforms, which must collaborate and communicate with each other. There are several advanced research platforms for wireless sensor networks in use by researchers globally with each platform offering unique differentiators such as sensor size, power consumption, nature of operating system, or basic sensing abilities [10,9]. In this chapter, we examine some sensor platforms and their associated software tools commonly in use today. Most tools used in the management of wireless sensor networks can be classified into three distinct categories (see also Fig. 3.1):

- Sensor level
- Server level
- Client level

3.1 SENSOR LEVEL

This level consists of devices that measure some physical phenomena or quantity such as sound, motion, light intensity, or temperature and convert it into some quantifiable form that can then be read by devices or human observers. Each sensor's mote contains an onboard sensing, communication, power, and processing module that allows it to perform sensing tasks. In the following sections we discuss some of the most commonly used sensor platforms and their features that make them unique.

3.1.1 The Mica Family

The Mica family of sensors is one of the most common sensing platforms in use; it is supported by numerous operating systems and sensing modules, including TinyOS,

---

Fundamentals of Sensor Network Programming: Applications and Technology, By S. S. Iyengar, N. Parameshwaran, V. V. Phoha, N. Balakrishnan, and C. D. Okoye

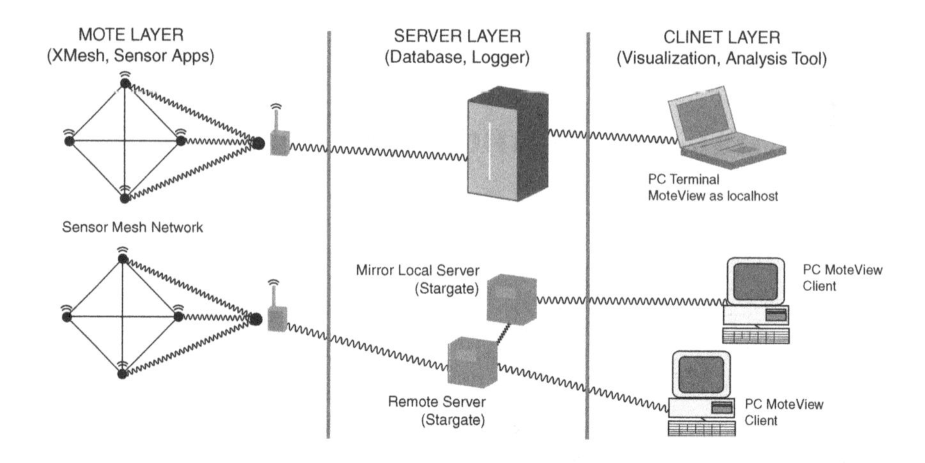

FIGURE 3.1 An illustation of the various distinct categories of sensor tools.

Mantis OS, and Contiki. It includes the MicaZ (Fig. 3.2), Mica2(Cricket) (Fig. 3.3), and Mica2Dot (Fig. 3.4) series of sensors.

Each of these motes has unique physical and functional capabilities, which serve as a key differentiator. Table 3.1 summarizes their similarities and differences effectively.

To extend the functionality of these motes, additional sensor data acquisition boards can be attached using the expansion connector provided on the Mica motes

FIGURE 3.2 A MicaZ sensor.

FIGURE 3.3 A Mica2 sensor board.

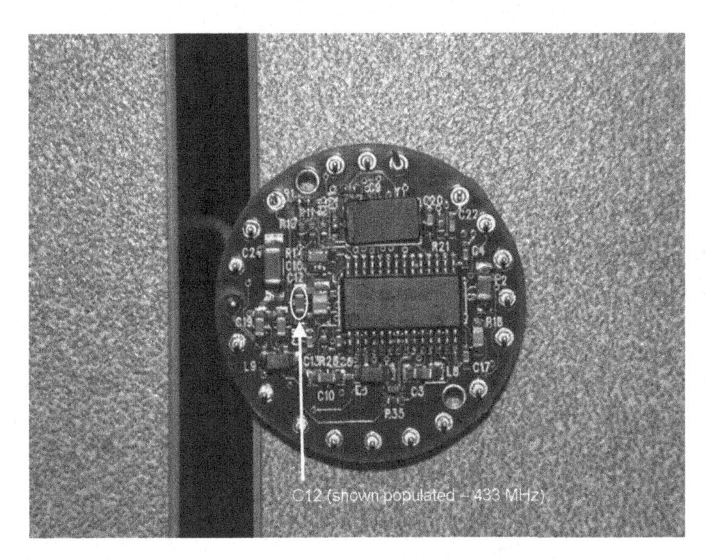

FIGURE 3.4 A Mica2Dot sensor.

TABLE 3.1 Mica Family of Sensors

Property	MicaZ	Mica2	Mica2Dot
Flash memory, kB	128	128	128
Measurement memory, kB	512	512	512
EEPROM, kB	4	4	4
A/D channels	10 bits (8)	10 bits (8)	10 bits (8)
Frequency, MHz	1400–2483.5	433/868/916	433/868/916
Data rate, kbps	250	19.2	19.2
Outdoor range, m	100	300	300
Size	6×3×1 cm	6×3×1 cm	2.5×0.6 cm

FIGURE 3.5 MTS300 data aquisition board.

and a general-purpose interface on the Mica2Dot motes. Figure 3.5 shows a typical sensor data acquisition board. Some other acquisition boards include the MTS300, MTS310 (Fig. 3.6), MTS101, and MDA500.

3.1.2 The Telos–Tmote Sky Family

The Telos family of sensors consists of TelosA and TelosB motes [5]. They are a newer generation of motes when compared with the Mica family, as they have a universal serial bus (USB) interface for data collection and programming.

In a similar design, the Tmote sky sensors contain a USB port to facilitate programming and are an exact replica of the TelosB suite of sensors. These features make them well suited for wireless sensor network experimentation in the research community. Figures 3.7 and 3.8 show the TelosB and Tmote sky sensors.

FIGURE 3.6 MTS310 data aquisition board.

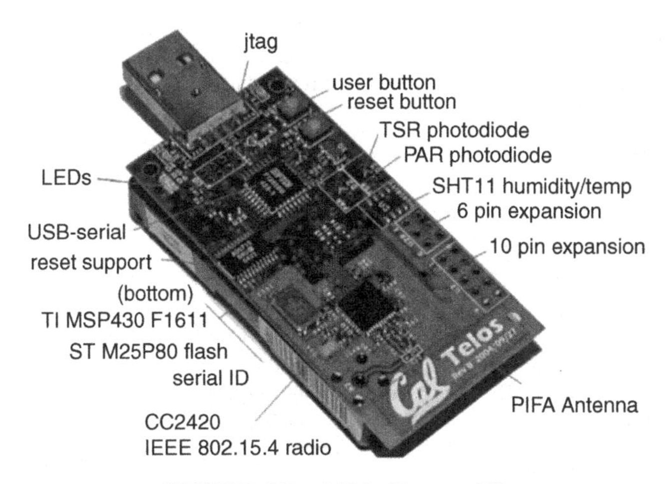

FIGURE 3.7 A TelosB sensor [8].

The functional and physical characteristics of both platforms are compared in Table 3.2.

As in the Mica family of sensors, additional data acquisition boards can be attached to these motes to allow for a more diverse sensing ability. One such example is the bumblebee radar board (Fig. 3.9), which provides a pulsed Doppler radar for TelosB and Tmote sky sensors.

3.1.3 Imote2

The Imote2 is a sensor platform built around the Intel PXA271 Xscale processor with a built-in 2.4-GHz antenna [1]. It is a powerful platform supporting computationally intensive tasks such as digital image processing, due mostly to the scaling capabilities

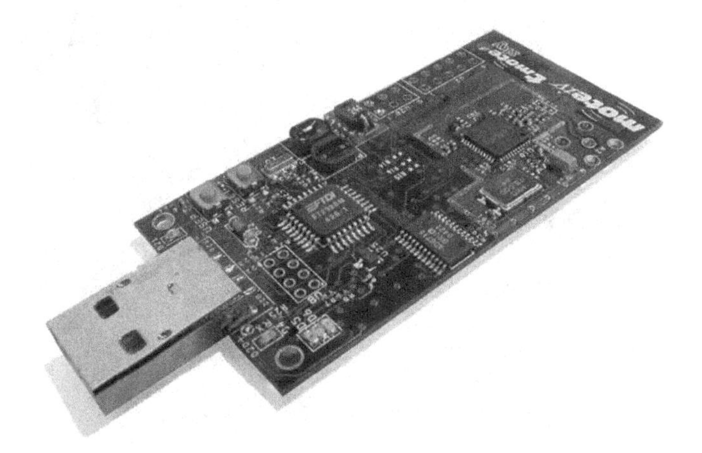

FIGURE 3.8 A Tmote sky sensor.

TABLE 3.2 Telos–Tmote Properties

Specifications	Telos Platform	Tmote Sky
Program flash memory, kB	48	48
RAM, kB	10	10
ROM, kB	16	16
A/D converter, bits	12	12
Frequency band, MHz	2400–2483.5	2400–2483.5
Data transmit rate, kbps	250	250
Outdoor range, m	75–100	50–125
Size, mm	65×31×6	65×31×6
Light-sensing range, nm	320–720	320–720
Temperature range, °C	−40–123.8	−40–123.8
Humidity range, % RH[a]	0–100	0–100

[a]Percent relative humidity.

of its processor, ranging from 13 to ~416 MHz. It has ~32 MB of memory and supports data rates of ≤250 kbps. It is currently supported by TinyOS and some Linux variants and can be ordered with the Microsoft.net microframework preinstalled. Figure 3.10 and Table 3.3 provide more information about the functional and physical characteristics of the sensor platform.

3.1.4 SHIMMER

SHIMMER, which represents sensing health with intelligence, modularity, mobility, and experimental reusability, is a sensor platform for health-related technologies. It

FIGURE 3.9 A TelosB sensor attached to the bumblebee radar board [7].

FIGURE 3.10 An Imote2 sensor.

supports wearable applications such as capture of real-time kinematic motion and physiological sensing. SHIMMER motes are driven by TinyOS and support up to 2 GB (gigabytes) of data storage for offline data capture (microSD storage). Some applications of this platform include sleep studies, cognitive awareness, vital signs monitoring, and chronic disease management. (see Table 3.4 and Fig. 3.11.)

Apart from the few platforms briefly covered in this chapter, there exist several equally important sensor platforms not covered in this book such as CSIRO's fleck platform for environmental monitoring, SNoW5 platform, and many others.

3.2 SERVER LEVEL

Gateways and programming boards that handle the buffering of data from the wireless network constitutes most devices at this level. Most programming boards are dual-purpose devices allowing direct access to motes for in-system programming and at the same time serve as gateways for communication with an existing sensor

TABLE 3.3 Imote2 Specifications

SDRAM memory	32 MB
Flash memory	32 MB
Frequency band	2400–2483.5 MHz
Data rate	250 kbps
Range of sight	30 m
I/O ports	3 UART, 2 SPI, SDIO, GPIO
Size	36 × 48 × 9 mm
Operating systems	TinyOS, Microsoft.net Framework

TABLE 3.4 SHIMMER Properties

Memory	10 kB RAM, 48 kB ROM
A/D converter	12 bits
Storage	2 GB microSD
Communication	CC2420 radio and class 2 Bluetooth
Sensors	MEMS accelerometer
Operating life	Deep-sleep life $>$ 1 year
Form factor	1.75 × 0.8 × 0.5 in.

network. These devices may support remote access to a sensor network as in the case of Crossbow's MIB600 Ethernet interface board or basic base stations relaying data from the sensor network to computer over their USB connectivity. The purpose of gateway devices is to connect sensor nodes to existing Ethernet networks. Table 3.5 summarizes the features of two popular interface boards—MIB520 (Fig. 3.12) and MIB600 (Fig. 3.13)—and a fully configured gateway device, Stargate network [2,4].

Unlike the software that runs on devices in the sensor level, all programs running on the server class of devices are in an always-on mode, which ensures the timely translation, processing, and buffering of data that emanate from the wireless network. These programs form the link between isolated sensing motes and the traditional Ethernet network running on an Internet client. One such example is Crossbow's

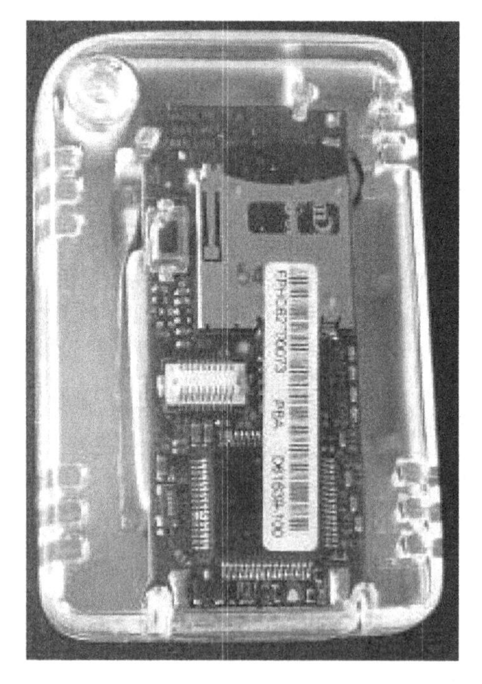

FIGURE 3.11 The SHIMMER platform.

TABLE 3.5 Gateway Device Properties

Feature	Stargate net-bridge	MIB520	MIB600
Program flash	8 MB	—	—
RAM	32 MB	—	—
I/O			
Extra features	1 × RJ45 and 2 × USB 2.0 USB flash disk 2 GB, onboard server	USB interface JTag interface for debugging	RJ45 POE, ARP, DHCP, Telnet
Size	130 × 21 × 91 mm		4.63 × 2.29 × 1 in.

FIGURE 3.12 The MIB520 programming board.

FIGURE 3.13 The MIB600 programming board.

XServe [3] software from their Moteworks development suite of applications based on TinyOS. XServe acts as a primary server running on a PC or gateway device from which the data gathered from the sensor network are interpreted. Other examples include the Global Sensor Network (GSN) middleware [6].

3.3 CLIENT LEVEL

The client level consists of all publishing, visualization, and monitoring applications. A number of powerful platforms exist today that support advanced publishing of sensor data. Global Sensor Network middleware supports publishing sensor streams to RSS feeds, security management system (SMS) text updates, Web publishing, and several other applications. Other examples include the Mote View from Crossbow [3].

3.4 PROGRAMMING TOOLS

Setting up and configuring your sensor programming environment can be a daunting task if there is in adequate knowledge of available resources. It is for this reason that we discuss how a TinyOS-based programming environment can be set up for both Linux and Windows operating systems.

3.4.1 Installing TinyOS in Linux

Using the very popular Ubuntu operating system, these guidelines explain how TinyOS can be installed and configured in a minimal number of steps. In administrator mode, perform the following steps:

1. In your fresh installation of Ubuntu, open the file `sources.list` located in `/etc/apt/` with your favorite text editor.
2. Add the TinyOS repository to your sources.list file

   ```
   deb http://tinyos.stanford.edu/tinyos/dists/ubuntu
   hardy main
   ```

3. Run the aptitude package management program to update local repositories:

   ```
   sudo apt-get update
   ```

4. Ensure that all necessary tools required are already installed

   ```
   sudo apt-get install build-essential
   ```

5. Install TinyOS:

   ```
   sudo apt-get install tinyos-2.1.0
   ```

6. On installing TinyOS, a few environment-ralated variables have to be set. In a Final step, edit your `.bashrc` file stored in your home directory and add the following lines:

```
export TOSROOT=/opt/tinyos −2.1.0
export TOSDIR=$TOSROOT/tos
export CLASSPATH=$TOSROOT/support/sdk/java/tinyos.jar
export MAKERULES=$TOSROOT/support/make/Makerules
export PATH=/opt/msp430/bin:$PATH
```

Your TinyOS installation and configuration are now complete. You can run the `check-env` command to view system sanity information.

3.4.2 Installing TinyOS in Windows

TinyOS is supported on the Windows platform through the use of Unix emulation software Cygwin. The following steps describe how TinyOS libraries can be installed on a Windows workstation:

1. Install the Java JDK. The latest version can be downloaded from the Sun Java Website
2. Download and install Cygwin from http://www.cygwin.com.
3. Install native compilers for sensor motes. The MSP430 toolchain for the Telos–Tmote family of motes or the AVR tool chain for the Mica family can be downloaded from http://www.tinyos.net/dist-2.0.0/tools/windows.
4. Install the nesC compiler from the TinyOS Website.
5. Finally, install the TinyOS source tree that will enable you to compile and install TinyOS programs. It can also be installed from the repository listed in step 4.

An alternative Windows installation method is to install the MoteWorks package provided free from Crossbow. Some more resources on Tinyos.net (available at http://www.tinyos.net/dist-2.1.0/tools/windows/) describe how TinyOS can be installed and configured in Redhat, Windows, and other Debian-based Linux distributions.

PROBLEMS

3.1 What are the three distinct categories of tools used for managing wireless sensor deployments?

3.2 Using the procedures outlined in the programming tools, install and configure your TinyOS programming environment.

3.3 What does the term `middleware` refer to?

3.4 Using the Global Sensor Network (GSN) reference provided, install and configure the GSN middleware.

3.5 What is the primary difference between server-tier tools and client-tier tools?

3.6 List three other examples of data-aggregating platforms such as GSN.

REFERENCES

1. *Crossbow Imote2 Datasheet,* courtesy Crossbow Inc.
2. *Crossbow MIB520 Datasheet,* Courtesy Crossbow Inc.
3. *Crossbow Moteworks Software Reference Manual,* courtesy Crossbow Inc.
4. *Crossbow Product Feature Reference Manual,* courtesy Crossbow Inc.
5. *Crossbow TelosB Datasheet,* courtesy Crossbow Inc.
6. *Global Sensor Networks,* GSNTeam.
7. http://blog.xbow.com/xblog/sensorboards.
8. http://inst.eecs.berkeley.edu/cs194-5/sp08/lab1/index.html.
9. Research Integration: Platform Survey, Embedded WiSeNts consortium.
10. M. Ruiz-Sandoval, T. Nagayama, and B. F. Spencer, Sensor development using Berkeley mote platform, *J. Earthquake Eng.* **10:**289–309 (2006).

PART II
Background

4 Data Structures for Sensor Computing*

Understanding the fundementals of data structures is essential to writing efficient code.
—S. S. Iyengar

Sensor computing consists simply in manipulation of sensor data in a suitably chosen data structure for event/data-driven sensor computing applications [5]. It can also refer to the design and analysis of sensor data abstractions. An in-depth understanding of the structural properties of certain data structures will yield very efficient algorithms for several important classes of problem. Because of the crucial importance of sensor data organization, this chapter deals with various data structures used in the design and analysis of sensor-based algorithms.

One of the basic facts that we need to understand is that sensor programming is different from traditional programming. The high-level coding in sensor programming is typical for any active object-oriented framework. Furthermore, the design is layered with a tiny operating system, the foundation for preemptive and roundrobin kernel, and other basic services such as event-driven and CPU pools. Figure 4.1 is essentially a design-based schematic diagram of program flow. Figure 4.1a is a schematic flowchart of a quickstart application, while Fig. 4.1b is a flowchart of an event-driven sensor application running on top of a cooperative vanilla kernel. At the highest level, the flowcharts are similar in that they both consist of a main loop surrounding various programming constructs. But the internal structure of the main loop is very different in the two cases. As indicated by the heavy lines in the flowcharts, the "quickstart" application spends most of its time in the tight "event loops" designed to busy–wait for receiving events, such as communication update events. In contrast, the "event-driven" application spends most of its time right in the main loop. The wireless sensor framework dispatches any available event to the appropriate state machine that handles the event and returns quickly to the main loop without ever waiting for events internally. A variation of event-driven programming is shown in Fig. 4.1c for battery-operated constrained wireless sensors. Here the data are transmitted periodically to a sensor's neighbors whenever new data become available and the battery level is

*Portions of this chapter are adopted from two earlier books by the author [2,7].

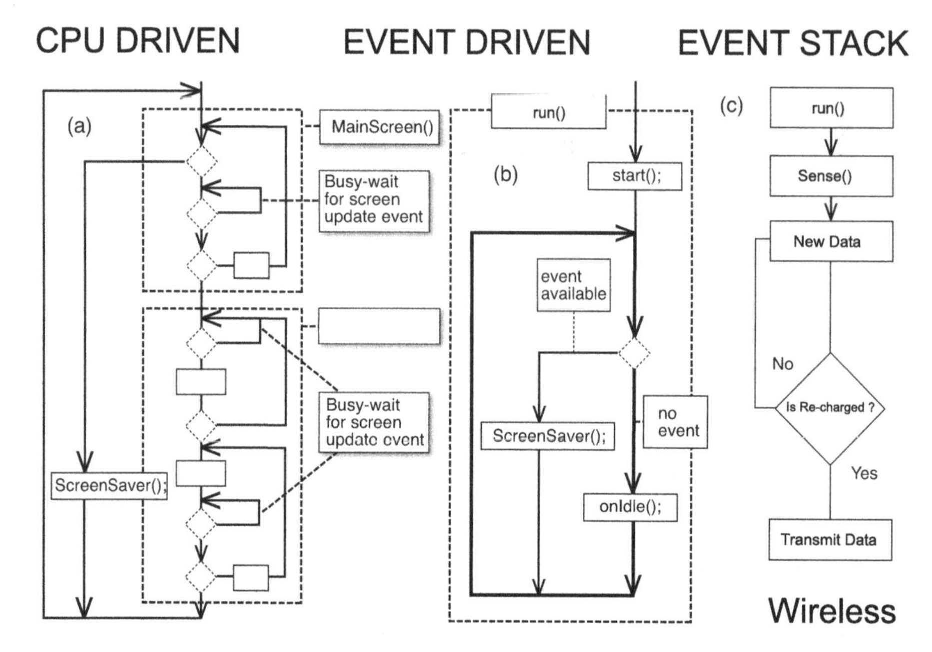

FIGURE 4.1 Control flow in (a) traditional, (b) event driven, and (c) wireless platforms.

above a given safety threshold. These broadcasted messages are received and queued by neighboring sensors and forwarded to the destination.

As a result of hardware constraints in sensor motes, we have presented the microframework memory usage in Table 4.1 (Table 4.1 shows the memory footprint of the components various settings for the configuration macros). The major difference between a fully functional node and a standalone node is that the latter does not have a kernel and acts like a data-sensing unit, whereas the fully functional node has a kernel and can route and receive messages efficiently. Therefore, a fully functional node has more RAM and ROM space.

TABLE 4.1 Resource Requirements For Target

Property	MicaZ	Mica2	Mica2Dot
Flash memory, kB	128	128	128
Measurement memory, kB	512	512	512
EEPROM, kB	4	4	4
A/D channels	10 bits (8)	10 bits (8)	10 bits (8)
Frequency, MHz	1400–2483.5	433/868/916	433/868/916
Data rate, kbps	250	19.2	19.2
Outdoor range, m	100	300	300
Size	6 × 3 × 1 cm	6 × 3 × 1 cm	2.5 × 0.6 cm

The application of all of these concepts of sensing/computing and communications are folded into a new sensor-based programming paradigm, which is the topic of the next section of this book.

4.1 INTRODUCTION TO SENSOR COMPUTING

The availability of effective communications, coupled with the computational capability of sensors, makes it feasible to host various tasks in sensors. As shown in Fig. 4.2, the four primary functions are the algorithm processing, process diagnostics, data management, and system interfaces. *Algorithm Processing* involves tasks that are performed in a sensor node. Specialized algorithms are required to condition signals, encrypt data, and process data in the node. Depending on the overall design of the distributed sensor network (DSN), the nodes may implement components of a distributed algorithm. The operating environment of a node is responsible for ensuring that these algorithms are executed fairly and effectively. *Process diagnostics* are additional computations that are performed at the sensor or cluster levels to augment the processing function of the input subsystem. Various techniques for automatically embedding code in the algorithms are being investigated (for diagnostics, monitoring, or distributed services). For example, such embedded code could provide status information and alarm data to operator monitoring stations. Some diagnostic strategies require temporal information in addition to the input data.

Data management is another function that is becoming increasingly important for DSNs. Because of the size of contemporary systems, the data gathered by the collection of sensors is immense. Typically, it is not feasible to associate mass storage devices at the level of a sensor, and the amount of memory available in a resource-constrained sensor is limited. Thus, it is necessary to manage the data in a DSN and effectively synthesize information that is useful for decisionmaking. Data management considerations for periodic systems are more critical because of issues of data freshness. The computing subsystem must support multiple system interfaces to effectively integrate with other systems. For the interface with the physical environment, it is necessary to interface to proprietary and open sensor interface standards. For example, several sensors interface with Ethernet or SERCOS. To allow users to work with emerging pervasive devices or to incorporate the DSN as an infrastructure for a smart space for automation [4], the DSN must support open interfaces that are based on XML (eXtensible Markup Language) or such other technologies.

The implementation of these functions is discussed under the categories of processing architecture, distributed services, and sensor operating systems [7,2]. Distributed services facilitate the coding and operation of a DSN and are provided by a distributed operating system that is represented by the collection of operating systems on each sensor. *Transparency* refers to the ability to regard the distributed system as a single computer. Tannenbaum [6] defines several forms of transparency for distributed systems: (1) data or program location, (2) data or process replication, (3) process migration, (4) concurrency, and (5) parallelism. For our purposes in a

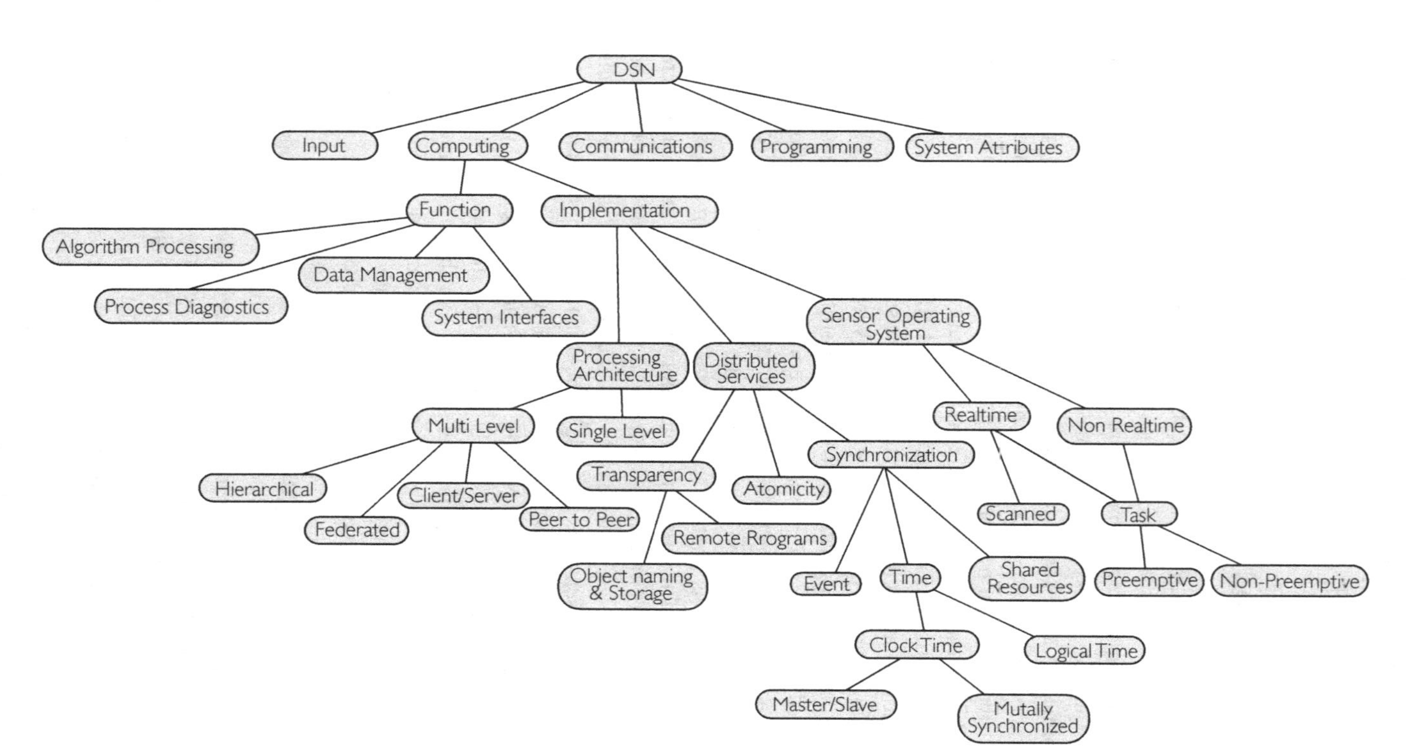

FIGURE 4.2 Taxonomy of a distributed sensor computing framework.

DSN, transparency concerns the *object naming–storage service*, which provides the ability to access system objects without regard to their physical location, and *remote program services*, which provide the ability to create, place, execute, or delete a program without regard to the sensor. Typically, servers are necessary to perform the registration and lookup functions to provide these services. The *atomicity service* is used to increase the reliability of the system by ensuring that certain operations (called *transactions*) occur in their entirety, or not at all. Various forms of recovery mechanism can be implemented to checkpoint and restore the component state should the atomic operation fail. Typically, atomicity is more important at the level of information-based transactions and less important at the level of periodic data gathering.

The order in which data from various sensors are gathered and the nature of interactions among the multiple sensors depends on the synchronization method. The *event service* allows a sensor to register an interest in particular events and to be notified when they occur. The time service is used to provide a systemwide notion of time. An important application of system time is in the diagnostic function, where it is used to establish event causality. Two forms of time are possible: clock time and logical time. Providing a systemwide clock time that is globally known within a specified accuracy to all the controllers in a distributed system can be difficult. Clock time can represent a standard Coordinated Universal Time (UTC), or it can be a common time local to the system. Two common techniques are (1) to provide a hierarchical master–slave system, in which the time in the "master" sensor device is transmitted to the other "slave" sensors; or (2) use a peer-to-peer distributed mechanism to exchange local times among various sensors. For certain applications, where affordable, it is possible to use global positioning system devices as master clocks (at the master sensor devices) to synchronize multiple controllers (at the slave devices) with the UTC. Logical time provides only the relative order of events in the system, not their absolute clock time. For many applications, exact time may not be as important as ensuring that actions occur in the correct sequence, or with in certain relative time intervals between events. Many algorithms can be rewritten to use logical time instead of clock time to perform their function. Providing logical clocks in a distributed system may be more cost-effective if the applications can be restructured. The management of shared resources across the network is supported through mechanisms that implement mutual-exclusion schemes for concurrent access to resources. All tasks in a sensor execute in an environment provided by the *sensor operating system*. This operating system provides services to manage resources, handle interrupts, and schedule tasks for execution. The operating system is said to provide real-time services if the length of time required for performing tasks is bounded and predictable. The operating system is said to be non-real-time if such services are not supported. Real-time services are supported by providing either a periodic execution model or a real-time scheduler (e.g., rate monotonic scheduling). These schedulers are priority-based and can be preemptive (interruptible) or not. Preemptive scheduling can provide the fastest response times, but there is an additional context swap overhead. Depending on the way in which the scheduler operates, the methods used to code computing, and the interaction with the communication interfaces, the execution in a sensor can be deterministic, quasideterministic, or nondeterministic. One of the main challenges in DSN research is to design efficient deterministic and quasideterministic sensor nodes.

4.2 COMMUNICATION CAPABILITIES

The communication subsystem is the primary infrastructure on which the DSN is constructed, and hence design choices made in this subsystem strongly affect the other capabilities of the DSN. Figure 4.3 presents a taxonomy of this subsystem. The primary functions in this aspect are data transport and bridging. We distinguish between three types of data, each having different characteristics. Input data gathered by sensors are typically limited to a few bytes and need guaranteed, deterministic message delivery to maintain integrity. Sensors communicate primarily to synchronize and to recover from failures. Thus, intersensor traffic is likely to be sporadic, contain more information (aggregated data), and be more suitable to quasideterministic or nondeterministic delivery mechanisms. *System data* refers to all the other data delivery needs that may or may not have hard real-time requirements. For example, data required for system monitoring and status alarms may be critical and real-time, whereas data used by Internet-based supervisory systems may not. Non-real-time system data, such as downloads, can typically be handled in a background mode using a "best effort" protocol. The bridging function, which transports data between multiple networks, is important in contemporary distributed systems such as DSNs that are likely to be integrated into existing engineering systems. *Bridging* refers to tasks performed on interface devices that connect two (or more) networks. The protocol used on the networks may or may not be the same. These intelligent devices provide services such as data filtering, data fusion, alternate routing, and broadcasting and serve to partition the system into logical subsets.

A communication protocol definition, such as that in the open systems interconnection (OSI), is designed as layers of services from low-level physical implementation, to media access, through networking, up to the application layer. Such layered communication protocols are unlikely to be implemented in resource constrained sensor nodes. For this taxonomy, we focus only on the media access communication (MAC) layer since it appears to be the layer where most variations occur. Under the MAC protocol implementation attributes we consider two attributes: the addressing scheme and the access mechanism. The method of addressing messages, called the *addressing scheme*, can be source-based, in which only the producing device's address is used in messages versus using the destination address to route the message. Source-based schemes can be extended to use content-based addressing, in which codes are used to identify the type of data within the message. Source- or content-based schemes are typically used on a broadcast bus, a ring, a data server, or when routing schemes can be a priori specified. Destination-based schemes are used when there is usually one destination or when the routing is constructed dynamically. The capability to provide deterministic service is strongly affected by the access method that establishes the rules for sharing the common communication medium.

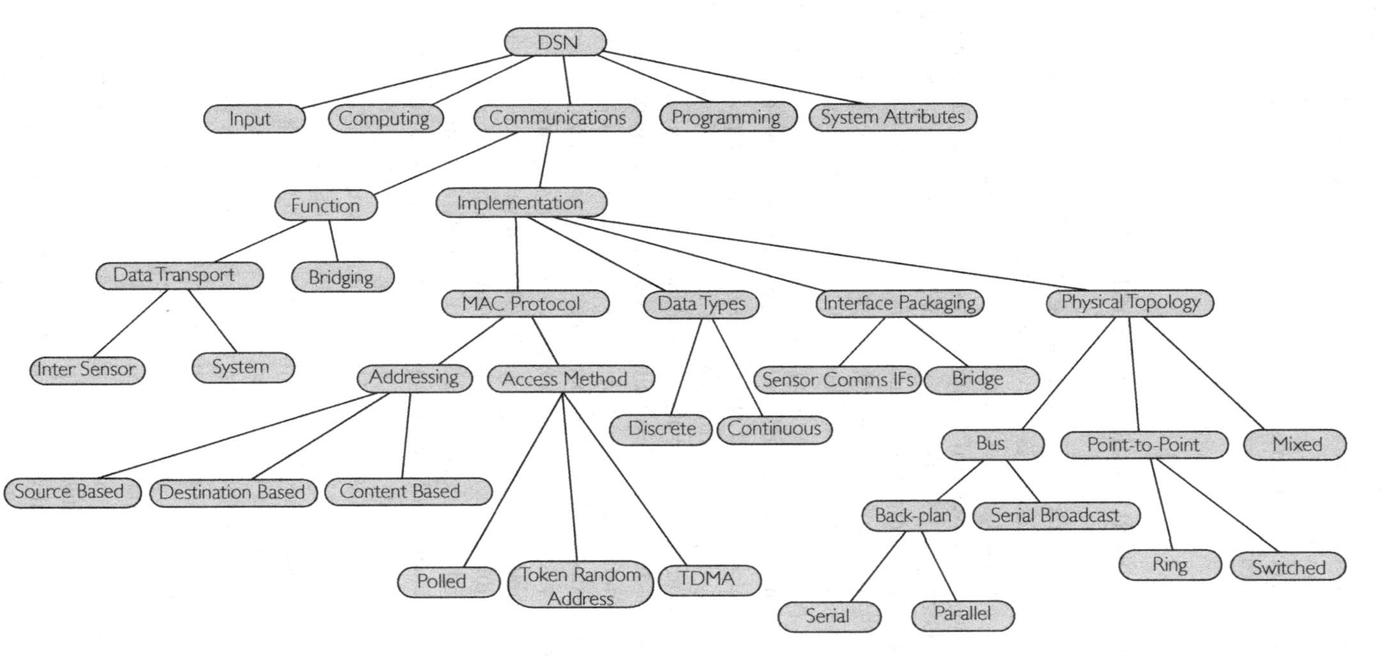

FIGURE 4.3 A taxonomy of a distributed sensor communications framework.

Polled, token-based, and time-division multiple access (TDMA) schemes that use a fixed timeslot allocation are deterministic. Token schemes, which allow nodes to skip their timeslot when they have nothing to transmit, have quasideterministic behavior. Random-access schemes, such as Ethernet, result in nondeterministic performance, and a priority bus scheme [e.g., controller area network (CAN)] can be made quasideterministic.

4.3 GENERAL STRUCTURE OF PROGRAMMING

This aspect has been largely ignored in the DSN literature. It must cover a range of activities, including designing, developing, debugging, and maintaining programs that perform computing, input, and communication tasks at the sensor level. Programs must also be developed to support distributed services that are essential for proper functioning of the DSN. In addition, activities such as abnormal-state recovery, alarming, and diagnostics must be supported. Figure 4.4 shows the primary functions of the programming category: support for coding of the algorithm, system testing, diagnostics, exception handling, data management, documentation, and synchronization. A key component of each function is the differences that are imposed by having to run in a distributed environment and which services are provided by the programming language and operating system. For example, the algorithm at a given sensor may require data from another sensor. An issue is whether the data are easily available (transparent services) or whether the programmer must provide code for accessing the remote data explicitly. System testing, diagnostics, and exception handling are complicated by the fact that data are distributed and determination of the true system state is difficult. Documentation includes the program source code and details of system operation. Questions of where programs and documents reside in the distributed system arise, as do issues in version control and concurrent access. Finally, the degree of transparency in synchronization that is provided by the languages and environment is a key to simplifying distributed programming. The language chosen in a DSN to implement the algorithm affects the services and tools that must provide support (e.g., operating system, compilers, partitioning, performance estimation). The IEC 1131 programming standards for digital controllers and the more recent IEC 61499 extensions that define an event-driven execution model are interesting considerations for programming DSNs. Ladder logic is relatively simple to learn, is easy to use, and provides a low-level ability to react to process changes. Sequential function charts, Petri nets, and finite-state machines (FSMs) are examples of state-based languages. An FSM model is intuitively simple, but the size of the model grows rapidly as the size of the control system increases. Hierarchical representation methods, such as hierarchical FSMs, have been used to cope with the large size of state-based models. While such hierarchical methods were well suited for hardware design, their use in software design is still an ongoing research issue. Function blocks are designed as a replacement for ladder logic programming in an industrial environment. They provide a graphical, software-IC (integrated-circuit)-style language that is simple to use and understand. Function

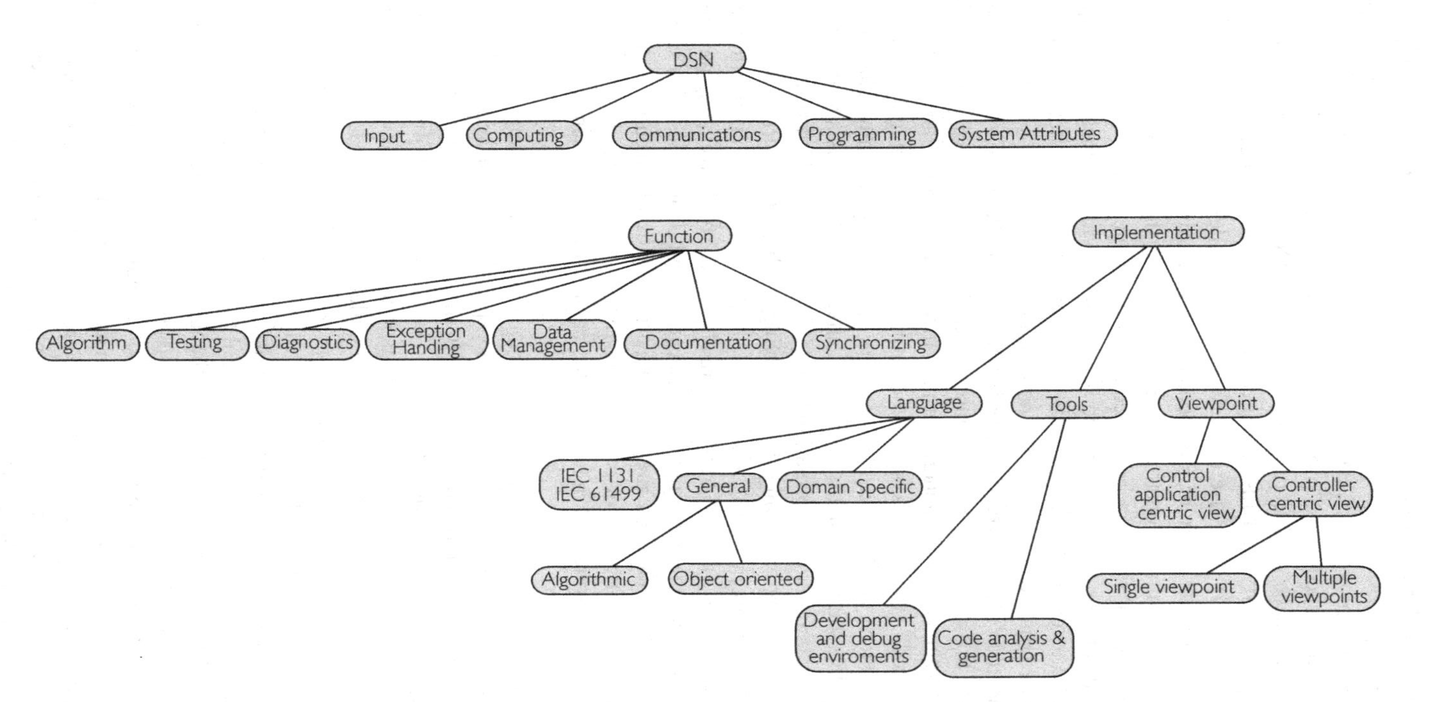

FIGURE 4.4 A taxonomy of a distributed sensor programming framework.

blocks are modular software units, can contain internal states, and represent the inputs and outputs of the function. Libraries of common blocks are provided for building control programs. General-purpose languages, such as FORTRAN or C, are employed to specify the processing in sensors. More recently, object-oriented languages have been used to program controllers. Domain-specific languages, with extensions to specify data fusion functions tailored to the needs of particular applications, are likely to be useful. Development and debugging environments for a DSN should support modular, independent programming of different sensors. Key distributed programming constructs can be provided to the programmer by distributed system services, or they can be embedded in the language and implemented by its compiler/linker. For example, a name server can provide the location transparently, or it can be a remote procedure call generated by the compiler. In addition, the programmer must be able to debug and maintain the system by viewing and manipulating the code in many sensors simultaneously. Formal models and theory help in simplifying this complex task. Because of the immense scale of DSNs, techniques that support the automatic generation and analysis of software are important. In an automated code generation system, the responsibility for managing and maintaining all the interactions between the sensors (by message passing, shared memory, or sharing I/O status) is handled automatically. Formal models and theory, such as Petri nets or compiler transformation theory, facilitate the task of software synthesis (and integration) by exploiting the underlying mathematical structure. The user is responsible only for providing a high-level specification of the application needs. In addition, the formal models and theory are also useful for introducing new functionality, such as abnormal-state recovery, alarming, and diagnostics. The viewpoint is another important issue in the programming aspect. Most of the current programming environments support a sensor-centric view. In this view, the needs of the control application must be expressed in terms of the capabilities of the sensor that is used in the DSN.

When dealing with large applications, managing such programs is a cumbersome activity. In an application-centric view, users express data fusion and integration needs by describing relationships among objects in the domain of the control application. Application-centric views can be supported with any level of abstraction (i.e., low, medium, or high). However, wireless sensor applications viewed with low-level abstraction tend to be more conducive to a collaborative approach (due to resource constraints) than would a traditional software program. In event-driven programming you deal with concepts such as inversion of control, blocking versus nonblocking code, and run-to-completion (RTC) execution semantics. Polling data are generated by the microframework, which allows global tracking of the current hardware clock. Once the poll is active, the sensor object changes states; if idle, it will accept the poll and initialize itself with a predetermined timeout. The timeout event is handled and the measured value from the sensor is read accurately and processed. If there is a threshold set for this particular process, then an alarm will be enabled according to the current read value. This alarm event will be broadcasted to reach its destination using source destination pairs. The "on idle" event can support the low-power features of the target hardware, enabling further energy savings.

4.4 DETAILS ON EMBEDDED DATA STRUCTURES

We now introduce the concept of developing active objects using a reusable infrastructure for specific domains such as sensor nodes and real-time systems. The infrastructure, an example of an application framework that is referred to as *microframework*, is a set of cooperating classes that makes up a reusable design. The microframework captures the overall architecture for executing concurrent sensor nodes in the embedded real-time environment.

The main element of decomposition of the microframework is an active object. An *active object* is a state machine that executes concurrently with other objects and communicates with them by sending and receiving events. This framework is typically compiled into the ROM of a senor node. The nodes have reprogrammable RAM area that the application uses for running and for storage.

Event queues and event pools are the necessary burden you need to accept when you work with active objects. The main problem with event queues and event pools is that they consume sensor nodes' precious memory. In order to minimize that memory usage, you need to size them appropriately. In this respect, event queues and pools are no different from execution stacks—these data structures all trade some memory for convenience of programming.

The correct sizing of event queues and event pools is especially important in microframework applications because the microframework offers no built-in handling for over flow or under flow of events in an event pool. These situations are both treated as bugs, no different from running out of execution stack space, with potential consequences that are just as disastrous.

Because of the crucial importance of data structures, this entire chapter deals with various data structures used in the design and analysis. For completeness, all the data structures are discussed. Those who are already familiar with the data structures may skip these topics.

An array usually represents a collection of homogeneous data items. An array is also a list. For example, the messages of a given sensor node are usually represented in the form of an array as shown in Table 4.2.

In the design and analysis of algorithms, we frequently come across two special types of arrays: stack and queue.

4.4.1 Stack

A *stack* is a one-dimensional data array in which addition and deletion take place from one end. Suppose that a stackcontains five data items. Then a designated variable top = 5 denotes that there are five items, where item 5 is the topmost entity. If anything

TABLE 4.2 Stack Items

Node ID	1	2	3	4	5	6	7	8	9	10
Received Messages	2	1	4	6	3	4	2	0	1	5

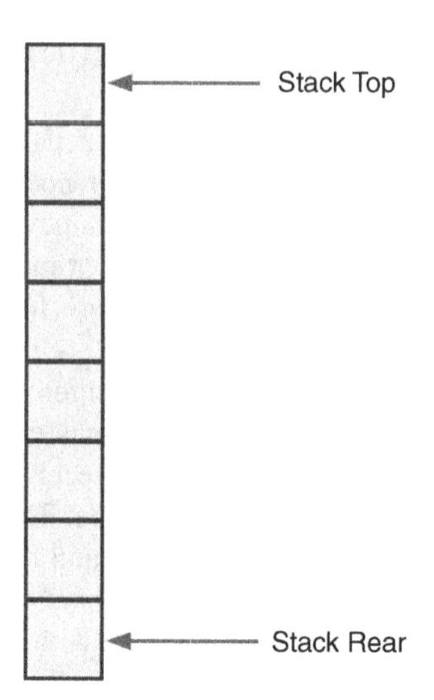

FIGURE 4.5 A stack with five entities.

new has to be added to the stack, it can be added as the sixth item, and the top will now become item 6. From a stack with five entities (see Fig. 4.5), if we want to delete one entity, we can remove only the topmost entity. The simple procedure for adding a new entity to a stack is shown below:

```
Procedure STACK-ADD(S(1:n), Top, item)
BEGIN
       IF top = n then
               call STACK-FULL
       ELSE
               top <-top+1
               S(top) <-item;
       Endif
END
```

```
Procedure STACK-DEL(S(1:n), top, item)
BEGIN
       IF top = 0 then
               call STACK-EMPTY
       ELSE
               item S(top)
               top <-top-1
       Endif
END
```

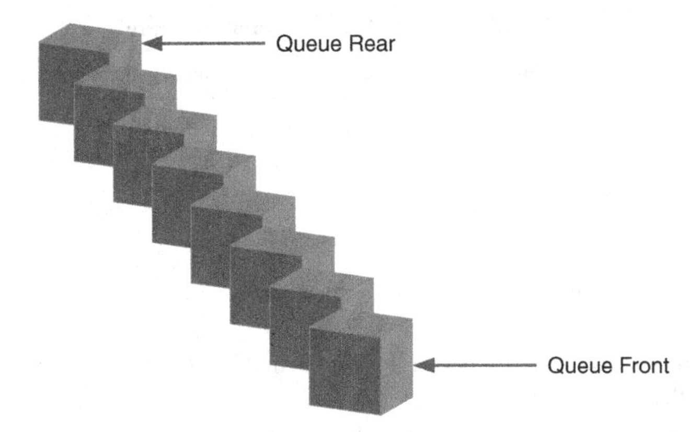

FIGURE 4.6 A queue.

In this procedure it is assumed that the maximum capacity of the stack is n. If the stack is already full, we cannot insert a new item in the stack. So, if top $= n$ we call a procedure "stack full," which (we assume) will eliminate the need for further processing. In a similar way we can write a procedure for deleting one item from the stack. If top $= 0$, we assume that the stack is already empty. In this case we cannot delete anything from the stack.

The stack data structure has numerous applications in algorithms—both sequential and parallel. The prefix/postfix/infix representation of arithmetic expressions, recursive procedures, scheduling, and polynomial evaluation are some important areas of application of the stack data structure.

4.4.2 Queue

The *queue* (see Fig. 4.6) is an array in which the additions take place through one end, called *rear*, and deletions take place through the other end, called *front*. In order to make the queue a convenient data structure for designing algorithms, it is represented in the form of a circular array. We assume that the queue is empty if and only if rear = front. It is easy to design procedures for insertion and deletions in a queue, and so it is left to the reader as an exercise.

4.5 LINKED LIST

The abstract data type called a *list* is a sequential collection of data items, called *atoms*, along with operations to work with the collection. If $a_1, a_2, \ldots, a_n$ are the atoms in a list, then we write the list as (a_1, a_2, a_n). A list differs from a set in that (1) there is an order to the items and (2) an item may appear more than once, if this is desirable in a given application.

TABLE 4.3 Linked List Data Structure

Name	Size
List 1	4 A, B, C, D
List 2	3 M, N, O
List 3	4 W, X, Y, Z

The common operations that one can perform on a list are

1. Find (item)—check whether the item is in the list and if so, indicate its position in the list.
2. Insert (item)—insert an item in the list (usually in a particular location).
3. Delete (item)—delete first (or possibly all) occurrences of the item.

The flexibility in the operation definitions is intended to accommodate different environments. If there is a way of comparing items a_i, and a_j, to say that $a_i < a_j$, or $a_i > a_j$, or $a_i = a_j$, then we may define sorted lists. An ascending sorted list is one where $a_i < a_{i+1}$ for all i. If a list is implemented by keeping the list atoms in an array, then we have a linear list. A typical linear list representation is in the form of an array $A(0{:}N)$, where $A(0)$ holds the number of items in the list. Such a representation requires limiting the number of entries to N and is clearly wasteful in the sense that $A(A(0)+1), \ldots, A(N)$ are unused. To keep a collection of several linear lists is awkward. Say, for example, that the lists (A,B,C,D), (M,N,O), and (W,X,Y,Z) are to be kept and the lists are allowed to grow to hold as many as 10 entries each. The lists may be regarded as a two-dimensional array $A(0{:}3,0{:}10)$. Table 4.3 describes the data structure of the lists described above.

Now consider holding all three lists in one array with an array of pointers into the array, shown in Table 4.4.

TABLE 4.4 Array of Three Linked Lists

Position	Data	Link
1	A	5
2	X	11
3	M	8
4	W	2
5	B	9
6	O	0
7	D	0
8	N	6
9	C	7
10	Z	0
11	Y	10

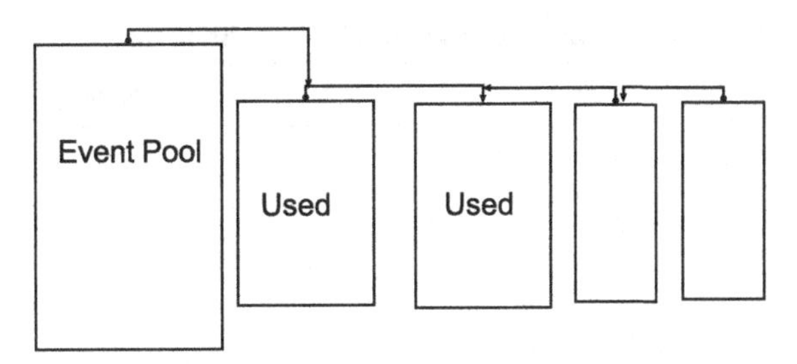

FIGURE 4.7 An event pool.

The array of pointers is called an *index*. This has the advantage of removing the upper bound on the list length; the only requirement is that the lists do not have a total length greater than the array size. Each list occupies only the space it needs and new lists may be easily added at the end (see Fig. 4.7). The problem here is, however, in inserting an item into the lists. The room to insert E in list 1 to make (A,B,C,D,E) can be obtained only by moving the rest of the data items in the list by appropriately changing the pointers. Another problem is that of deleting lists. It is awkward to keep track of free space within this large array. Thus an effective implementation for the maintenance of several lists is not to be found using linear lists.

4.5.1 Examples of Linked Lists

An alternative data structure to linear lists that allows for a greater amount of flexibility is the linked list. List elements are called *nodes*. A node consists of data and a link to another node in the list. The beginning of the list is indicated by a pointer. The name of the list is actually a pointer to the first element in the list. A linked list structure for our previous example is given in Table 4.6.

The pointer is set to 0 if it would take any value that cannot be used as a link, or if no link exists. If we extract the first, second, and third lists, we could draw it as shown in Fig. 4.8. The name of the list structure is *one*. In particular, the lists can grow to arbitrary length. We may be able to keep several lists simultaneously in one area of storage with the only space restriction that their total storage utilization will be less than or equal to the area available.

There is an obvious disadvantage to linked lists in terms of searching. If a linear list is kept in sorted order, then it may be searched with a binary search. Consider

TABLE 4.5 Linked List Data Structure

Name	Pointer
One	1
Two	5
Three	8

TABLE 4.6 Linked List Data Structure

Name	Pointer
One	1
Two	6
Three	9

the linear list *T* shown in Fig. 4.9a and the equivalent representation in linked lists shown in Fig. 4.9b. The binary search functions (see Table 4.7) can be obtained by computing midpoints of the list. The following discussion describes the binary search process. The midpoint between array elements 1 and 5 in the linear list given above is array element 3. But in the linked list there is no way to find the node midway between two other nodes, so a binary search is impossible.

The most obvious reason to use a linked list framework data structure such as the one described above is the ability to insert or delete dynamic events during the event detection process of a sensor network. This framework also frees application programmers from dealing with buffer-related communication issues and allows them to concentrate on robust methodologies for complex applications.

4.5.2 Circular Lists

In circular lists the pointer cell of the last node is not null. It points the first node. This is shown in Fig. 4.10a.

4.5.3 Doubly Linked List

The basic difficulty of a linked list or circular list is that we may move in only one direction. To move in both directions we introduce the concept of a doubly linked list. The nodes in a doubly linked list have two links—`LLINK` for the left link and `RLINK` for the right link—and every list has pointers to the left end and right end. Thus a typical doubly linked list resembles the structure shown in Fig. 4.10b. Note that there is one linked list found by following `RLINK` to a node and another list can be found by following `LLINK` to a node. It is particularly easy to delete a node in a linked list. The links allow you to find all different nodes. Suppose that a node `X` has left and right adjacent nodes.

1	2	3	4	5	6	7	8
A	B	C	D	M	N	O	W

FIGURE 4.8 Continuous list.

TABLE 4.7 Binary Search Example

A	*B*	*C*	*D*	*E*	*M*	*N*	*O*	*W*	*X*	*Y*	*Z*
1	2	3	4	5	6	7	8	9	10	11	12

Node X may be deleted by applying the following instructions:

```
RLINK(LLINK (X)->RLINK(X)
LLINK (RLINK(X)) = LLINK (X)
```

This is seen to produce the following doubly linked list structure, which effectively bypasses X. The two lists in the doubly linked list may themselves be circular (see Fig. 4.11). To insert a node containing a desired data item, we first must acquire an unused node. For this purpose we keep a special linked list called the *free list*, which holds nodes that are not currently used. Such a list may be initialized when memory is first allocated for nodes.

4.6 IMPORTANCE OF GRAPH CONCEPTS IN SENSOR PROGRAMMING

Knowledge and understanding of the fundamental properties of graphs is critical in distributed sensor programming. In this section we examine some of the ways in which knowledge of graphs greatly simplifies the task of sensor programming.

4.6.1 Network Localization

Sensors are typically deployed in an area to measure spatial and temporal characteristics of events of particular interest. This implies that these sensors need to convey data

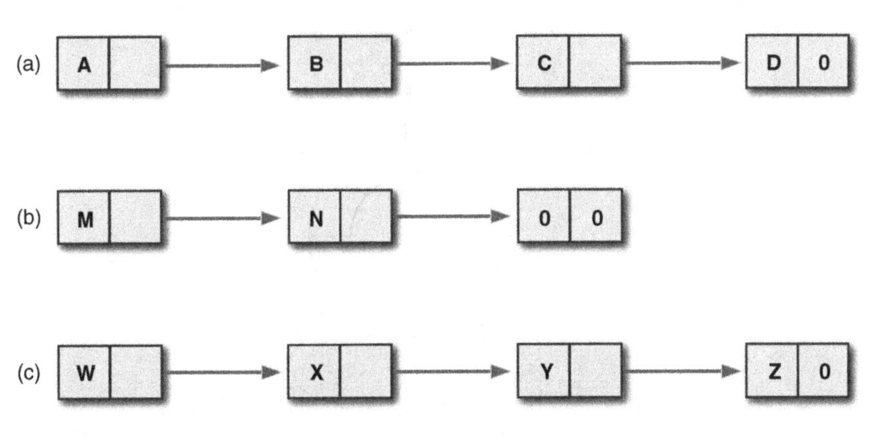

FIGURE 4.9 Three linked lists.

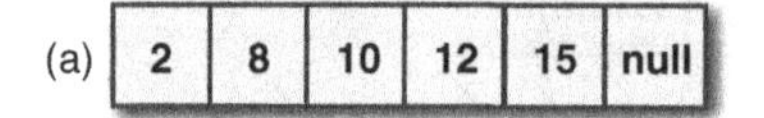

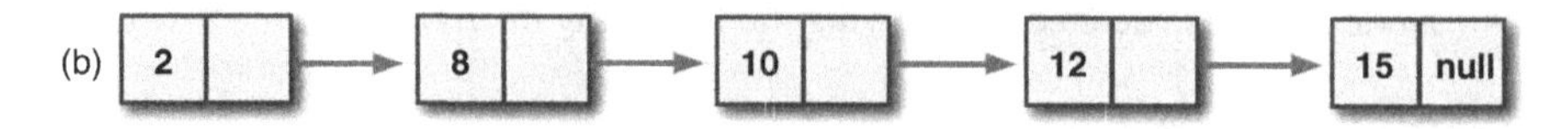

FIGURE 4.10 (a) A list; (b) a linked list.

about the event and also to relay information about the location of where that event occurred before any useful task can be done with that data. In network localization, knowledge of graphs can be useful in certain scenarios to provide an estimate of the distance between nodes in a network. Several node localization algorithms have been published that make use of concepts in graph theory.

4.6.2 Data Aggregation

The ability to aggregate data from distant nodes for more efficient transmission is an important data-processing primitive for sensor networks, particularly since it results in huge energy savings, extending the life of a deployed sensor network. For data aggregation to be successful, routing algorithms have to make intelligent decisions when choosing nodes to partake in data transmission from the source node to the data sink or otherwise risk depleting the energy of nodes in most frequent use. The most natural data structure that comes to mind when routing information based on associated costs and other parameters is a connected graph. It is for this reason that several tree-based data aggregation algorithms exist such as the *collection tree protocol* (CTP) in TinyOS.

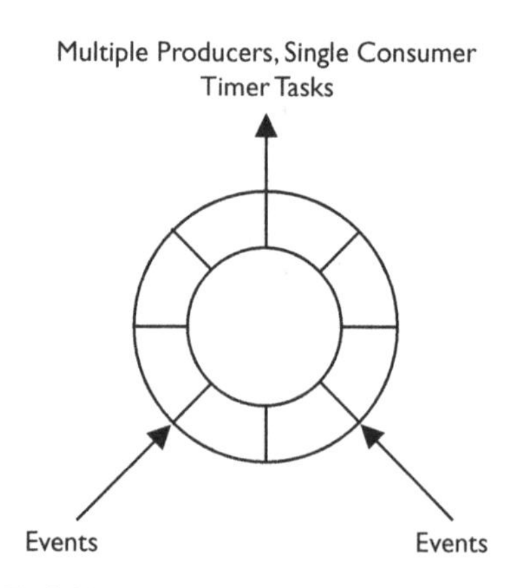

FIGURE 4.11 Circular linked list; doubly linked list.

4.6.3 Collaborative Processing

In most sensing tasks, sensors rarely rely solely on their own sensing abilities but rather distribute a sensing task among a subgroup of sensors and collaborate on that task. One example that comes to mind is any tracking task. Most entities being tracked are mobile in nature and for this reason, relevant sensors around the region of interest have to be selected dynamically for collaboration with other sensors on the basis of location and other factors. The task of choosing an appropriate sensor can be greatly simplified by applying concepts from graphs.

4.6.4 Planarity Testing

In graphs, the edges between nodes may intersect. This could be due to either of two factors: (1) the graph is nonplanar or (2) a planar graph is not drawn properly to exhibit its planar structure.

To ensure that a given graph is planar, a well-known computer science problem, several well-known practical algorithms have been proposed in the literature that run in $O(n)$ time where n is the number of vertices. One of them is *Kuratowski's theorem*, which states that a finite graph is planar if and only if it doesn't have a K_5 or $K_{3,3}$ as its subgraph.

Planarity testing is widely used in wireless sensor networks for problems associated with coverage, localization, routing, topology control, and detection of holes.

4.6.5 Graph-Coloring Concepts in MAC-Layer Protocols

The minimal graph-coloring problem is defined as the minimum number of different colors that are required to color the vertices of a graph such that no two adjacent vertices share the same color. This fundamental graph-theoretic problem is widely used in sensor networks for various applications. Ranging from designing efficient interference-free MAC-layer protocols service discovery, data aggregation (Fig. 4.13), TDMA slot assignments, and other scheduling assignments are approached using the graph-coloring problem. For example, in scheduling for wireless sensor networks, the individual nodes must be scheduled in such a manner that

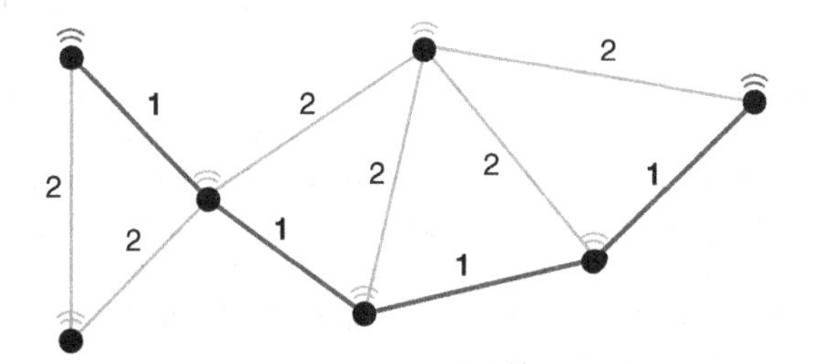

FIGURE 4.12 A simple illustration of node localization.

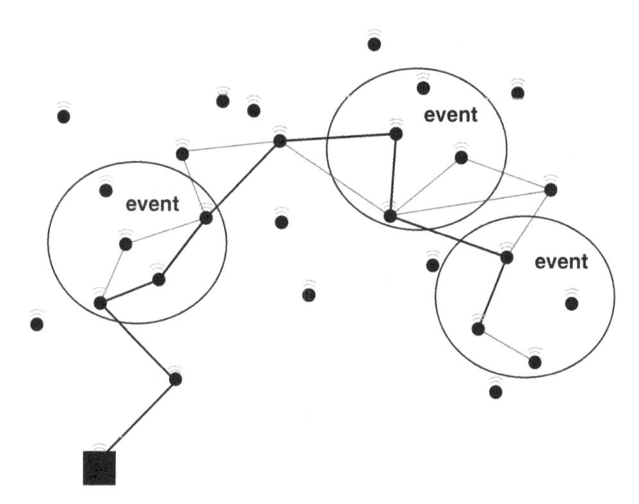

FIGURE 4.13 A simple illustration of data aggregation.

nearby nodes do not attempt to transmit data at the same time or should be scheduled in such a way that the nearby nodes use different frequency/channels for transmitting to avoid cochannel interference. By efficiently scheduling the nodes using graph coloring, one can minimize the energy consumption and latency in a given topology.

4.6.6 Isomorphism Applications in Sensor Deployment

In graph theory, two graphs G and H are said to be isomorphic to each other if there is a bijection between the vertex sets of G and H. In other words, although G and H may look different when drawn, they are essentially the same graph. One of these graphs (say, H) can be redrawn in such a way that it looks identical to G.

Inpractice, it is very difficult to identify isomorphism among given graphs. Many ad hoc deployments of sensor networks need localization techniques to discover and form a network. Several algorithms exist for efficient topology control and localization techniques for known types and structures of graphs. To identify or to restructure the deployment toplogy to a known structure, graph isomorphism will be useful.

In nesC programming, knowledge of fundamental data structures that preserves certain contexts also becomes necessary as real-world applications get written. For example, a queue or linked list can be implemented to store notifications or events in the order in which they occur. The importance of data structures cannot be over emphasized since efficiency and effective usage of resources are synonymous with sensor network applications. The use of inappropriate data structures can only lead to the rapid demise of existing sensor deployments, and for this reason, the next several sections will delve into the details of graphs.

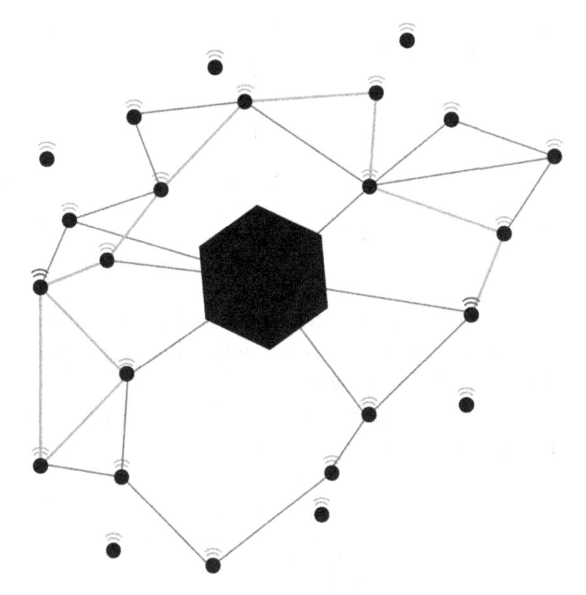

FIGURE 4.14 A simple illustration of collaborative processing among nodes.

4.7 GRAPH AND TREES

One of the most powerful ways to foster dynamical event-tracking applications is to use an embedded data structure such as a universal data structure. This data structure helps in developing a source of conceptual integrity from a programming perspective. This captures the dynamic connectivity of the sensor network at each node (see Fig. 4.14). As each node can be added to or deleted from the network in real time, it would need to have to have a unique address and information on its relative position. This mechanism allows us to dynamically maintain the node addresses consistently over time. This section introduces an important data structure, the graph model [3,1], and explains various types of graphs and some properties that are used in later chapters.

4.7.1 Preliminaries

Let V be any set. Let be a subset of $V \times V$. The pair (V, E) is called a *graph*. We denote $G = (V, E)$. V is called the *vertex set* and E the *edge set*. The element of V are called *vertices* and the elements of E are called *edges*. Let $V = a,b,c,d,e,f$ and $E = (a,b),(b,c),(c,d),(c,e),(a,d),(b,d)$. The graph $G = (V, E)$ can be represented by Fig. 4.15. If $e = (v_1, v_2)$ is an edge, then we say that e incidents on v and v_i. In such case v_i and V_2 are said to be adjacent vertices. An edge (v, v) is called a *self-edge*. If $e_1 = (v_1, v_2)$ and $e_2 = (v_1, v_2)$, then e_i and e_j are said to be *parallel edges*. A graph having no parallel edge and no self-edge is called a *simple graph*. If the vertex set is an infinite set, then the graph is called an *infinite graph*. Otherwise the graph is said to be a *finite graph*. A graph may have its edge set to be empty. Such a graph is

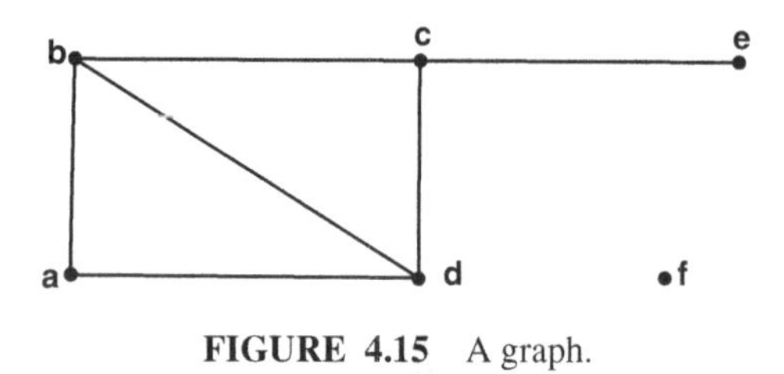

FIGURE 4.15 A graph.

called a *null graph*. A vertex in which no edge incidents is called an *isolated vertex*. The number of edges incident on a vertex v is called the degree of v. In the graph represented in Fig. 4.15, f is an isolated vertex.

The degrees of the vertices are given below. A vertex whose degree is 1 is called a *pendant vertex*. In our

Vertex a b c d e f

Degree 2 3 3 3 1 0

graph e is a pendant vertex. We usually denote the number of vertices and the number of edges by n and m, respectively. We are interested in the sum of the degrees of all the vertices. Each edge incidents on two vertices; so the presence of each edge contributes 2 to sum of the degrees. Hence we have the following observations:

1. The sum of the degrees of all the vertices is $2m$.
2. The number of vertices of odd degree is alway seven.

4.7.2 Regular and Complete Graphs

A graph in which all the vertices are of equal degree is called a *regular graph*. If a simple graph has n vertices, at the most $n(n - l)/2$ edges are possible. A graph with all possible edges is called a *complete graph*. Let K denote a complete graph with n vertices [and $n(n - l)/2$ edges]. Note that K is a regular graph of degree $n - 1$. This is illustrated in Figs. 4.16 and 4.17.

4.7.3 Walk, Path, Cycle

Let $G = (V, E)$ be a simple graph. Let $v_1, v_2, v_3, v_4 \ldots, v_k$ be some vertices of G, and let v_i, be adjacent to $v_i + 1(l = i = k)$. We say that this sequence $v, v_2, \ldots, v_i$ is a walk from v_i to v_k if no edge appears more than once in the sequence. v is called the *starting point* and u^* is called the *terminus*. In the preceding definition of a walk, some v_i and V_j may be the same for distinct i and j. Consider a walk $v_1, v_2, \ldots, v_k$. We say that the walk starts from v, travels through $v_1, v_2, \ldots, v_k$, and finally reaches v_k. If $v_1 = v_k$, the walk is called a *closed* walk. A walk that is not closed is said to be *open*. A walk in which no vertex appears more than once is called a *path*. A closed

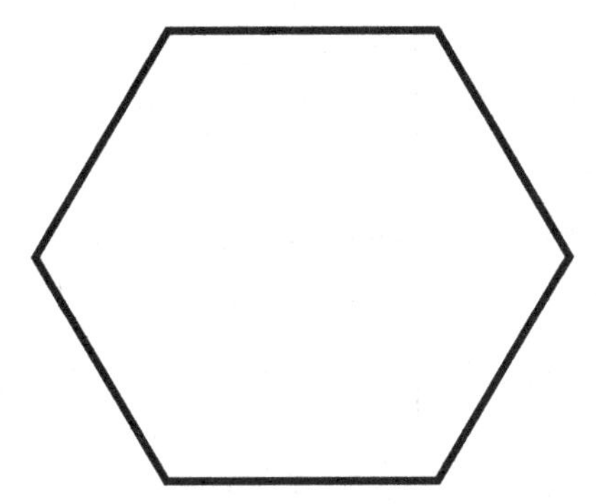

FIGURE 4.16 A regular graph of degree 2.

walk in which no vertex appears more than once is called a *cycle* or *circuit*. In other words, a circuit is a closed path.

In Fig. 4.18, a,b,c,d,b,e,f, is a walk. It is not a path because b is repeated; a,b,c,d, is a path from a to d. Also note that c,b,e,p,h,e,b,d is neither a path nor a walk. It is not a walk because the edge (b,e) appears twice in the sequence: once from b to e and then from e to b. The sequence a,b,e,f,g, a is a cycle. In a cycle an edge joining two nonconsecutive vertices is called a *chord*. In the cycle a,b,e,f,a, in Fig. 4.18, (b,f) is a chord of the cycle a,b,e,f,g,a.

4.7.4 Subgraph

Let $G = (V, E)$ be a graph. Let V' be a subset of V and E' a subset of E such that all the edges in E' incident only on vertices of V. $G' = (V',E')$ is called a *subset* of G. The graph shown in Fig. 4.19b is a subgraph of the graph in Fig. 4.19a. In a graph G, for any two vertices u and v, if there is a path from u to v, the graph G is called a *connected graph*. In a *disconnected graph*, a maximal connected subgraph is called a *connected component* or simply a *component*. Let V be a subset of V. Let be the collection of all edges of G, which have both end vertices in V. Then

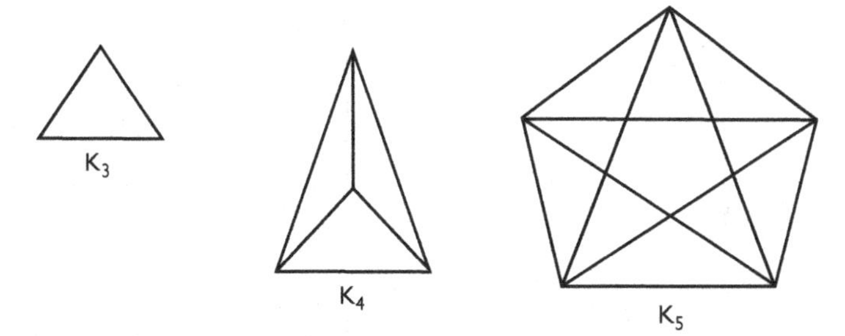

FIGURE 4.17 Some completed graphs.

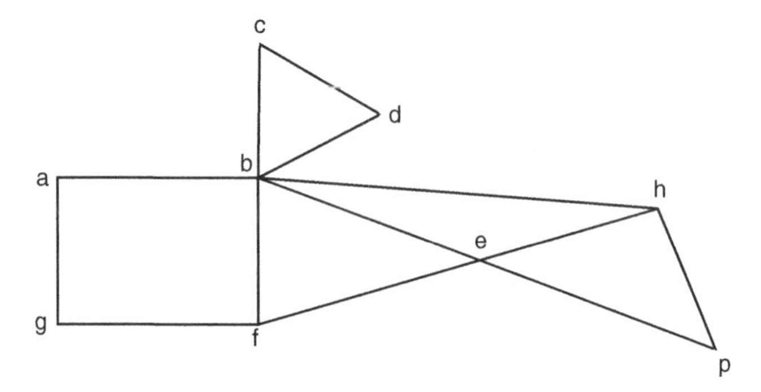

FIGURE 4.18 A graph illustrating paths and walks.

$G' = (V, G)$ is called the *induced subgraph* of G induced by V. The graph in Fig. 4.19b is not an induced subgraph of Fig. 4.19a. The graphs in Figs. 4.19c and 4.19d are induced subgraph, induced by 1,5,4 and 2,3,4, respectively. Two subgraphs are said to be *edge-disjoint* if they have no common edge. The graphs shown in Figs.

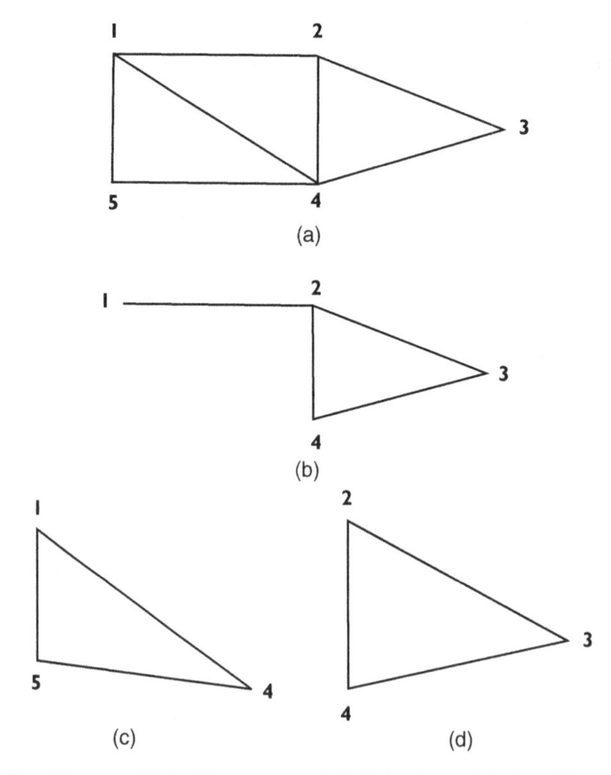

FIGURE 4.19 (a) A graph G; (b) a subgraph G; (c) subgraph induced by 1,4,5; subgraph induced by 2,3,4.

4.19c and 4.19d are edge-disjoint subgraphs of Fig. 4.19a. The following observations can now be made:

1. Any graph G is a subgraph of G itself.
2. Any vertex in the graph can be considered as an induced subgraph of the graph with one vertex.
3. Any edge (u, v) can be considered as an induced subgraph induced by u, v.
4. The relation subgraph of is reflexive, antisymmetric, and transitive.

Unlike simulated conditions of sensor deployments, which typically assume ideal conditions for sensors, real-world deployments are usually faced with a barrage of issues stemming from unanticipated scenarios such as the effects of the environment/landscape on sensor communication, or even interference by other RF-emitting devices. To handle such issues, communication routines in sensor networks must possess dynamic capabilities, computing alternative communication routes when disruptions occur, and as such, rely on critical concepts discussed in the following sections. Sections 4.7.5 and 4.7.6 introduce the topics of homeomorphism and isomorphism in terms of localization and routing in sensor networks.

4.7.5 Homeomorphism

Let G be a graph. Let v_1, v_2, v_j, be a path and let the degree of U_2 be 2 in G. The pairs (v_1, v_2) and (v_2, v_3) are called *series edges*. The operation of replacing the two edges (v_1, v_2), and (v_2, v_3), by a single edge (v_1, v_3) and removing the vertex V_2 is called the *merging* of series edges. If (u, v) is an edge, the operation of introducing a new vertex w and making it adjacent to both u and v and then removing the original edge (u, v) is called insertion of a vertex of degree 2. Two graphs, G_1 and G_2, are said to be homeomorphic if one can be obtained from the other using a finite number of operations merging of series edges and/or insertion of a vertex of degree 2. A complete subgraph is called a *clique*. A clique that is not a subgraph of any other clique is called a *maximal clique*. Consider the graph shown in Fig. 4.20a. Its maximal clique is shown in Fig. 4.20b.

4.7.6 Isomorphism

Two graphs $G' = (V', E')$ and $G' = (V, G)$ are said to be *isomorphic* if there is a bijection between V and V' in such away that two vertices of V are adjacent if and only if the corresponding vertices in V' are adjacent. If two graphs are isomorphic to each other, then they have an equal number of (1) vertices, (2) edges, and (3) vertices with the given degree. Note that these three conditions are necessary but not sufficient. It is an interesting research problem to find a simple and efficient criterion to check whether two given graphs are isomorphic. This problem is called the *isomorphism problem*.

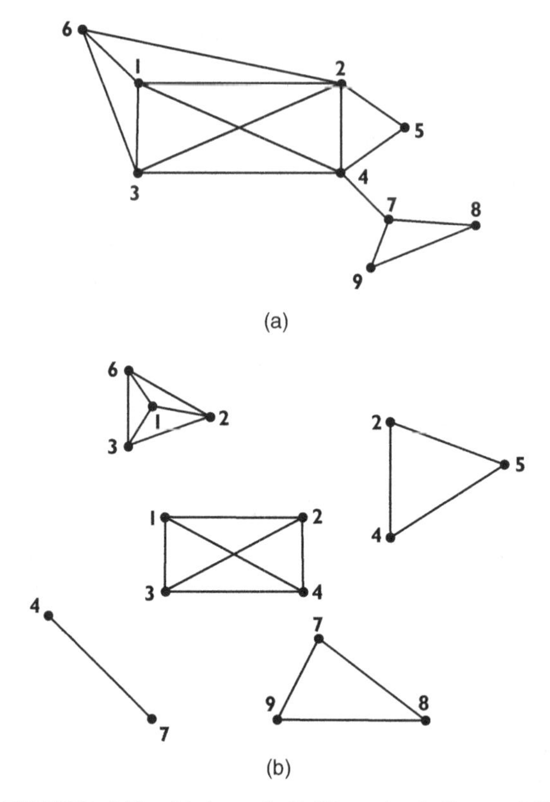

FIGURE 4.20 (a) A graph G; (b) maximal cliques of G.

4.8 TREES

A connected graph having no cycles is called a *tree*. The following statements are equivalent:

1. G is a tree.
2. There is exactly one path between any two vertices of G.
3. G is connected and contains n vertices and $n - 1$ edges.
4. G is minimally connected.
5. G has no cycles and G has n vertices and $n - 1$ edges.

Any tree has at least two pendant vertices. The length of a path is the number of edges that it has. The length of the longest path from a vertex v to any other vertex is called the *eccentricity* or *diameter* of v and is denoted by $E(v)$.

$$E(v) = \max d(v, u)/u \in V, \tag{4.1}$$

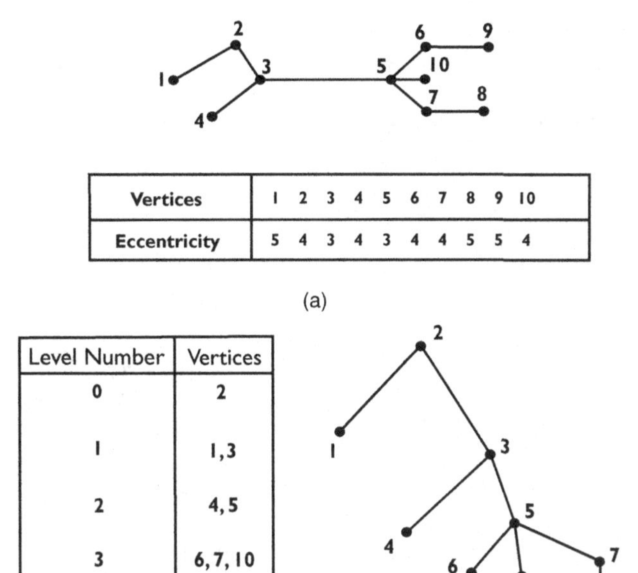

Vertices	1	2	3	4	5	6	7	8	9	10
Eccentricity	5	4	3	4	3	4	4	5	5	4

(a)

Level Number	Vertices
0	2
1	1,3
2	4,5
3	6,7,10
4	8,9

(b)

FIGURE 4.21 (a) Tree with radius = 3 and diameter = 5; (b) a rooted tree.

where $d(v, u)$ represents the length of the path from v to u. In a tree T, the vertex having minimum eccentricity is called a "v to u center of the tree." A tree contains one or two centers. In Fig. 4.21, the eccentricities of all the vertices are given. Points 3 and 5 are the vertices that have the minimum eccentricity. So, this graph has two centers, 3 and 5. The eccentricity of the center is called the *radius* of the tree. The length of the longest path in the tree is called the *diameter* of the tree. For the tree given in Fig. 4.21a, the radius is 3 and the diameter is 5. We can designate a vertex of the tree as its root. If a vertex is designated as the root of the tree, the tree is called a *rooted tree*.

Consider the tree given in Fig. 4.21a. If we designate 2 as the root of the tree, the tree can be redrawn, as in Fig. 4.21b. For a rooted tree, level numbers can be defined to each of the vertices as follows: The root is assigned level number 0. The vertices adjacent to the root are called the "children" of the root, and they are assigned the level number 1. The root is called the "parent" of its children.

In our tree 1 and 3 are the children of 2. If a vertex v is at level i, then any other adjacent vertex u that is not the parent of v is assigned level $i + 1$. Such a node will be called a "child" of v. If the maximum level number of a rooted tree is k, then its height or depth is defined as $k + 1$. The tree in Fig. 4.21b is of height 5. A connected subgraph of a tree is called a *subtree*. Figure 4.22 shows a tree and some of its subtrees. A rooted tree is usually represented by its parent relation. If $T = (V, E)$

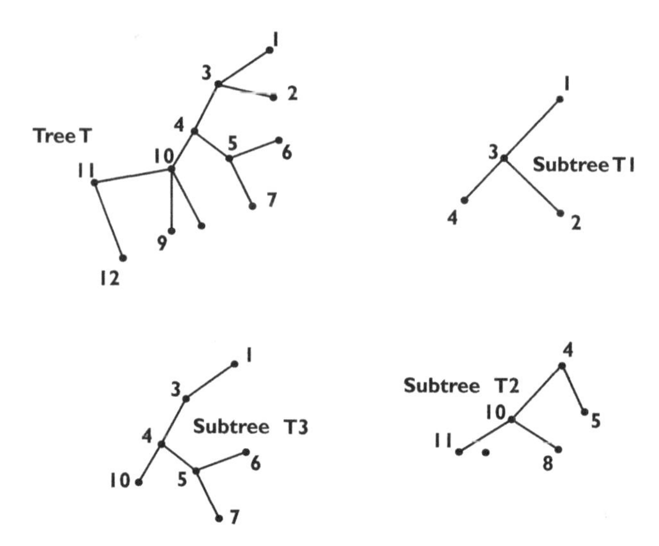

FIGURE 4.22 A tree and someof its subtrees.

is a rooted tree with root r, it is represented by the array parent $(1 : n)$ (where n is the number of vertices) defined by

```
PARENT(i) = the parent of i, if i is not the root.
PARENT(r) = r, where, r is the root.
```

In some cases the parent of the root is defined as -1. A tree and its parent representation are shown in Fig. 4.23.

4.8.1 Binary Trees

A rooted tree in which every vertex has at most two children is called a *binary tree*. Binary trees are widely used in computer science applications. The two children of a

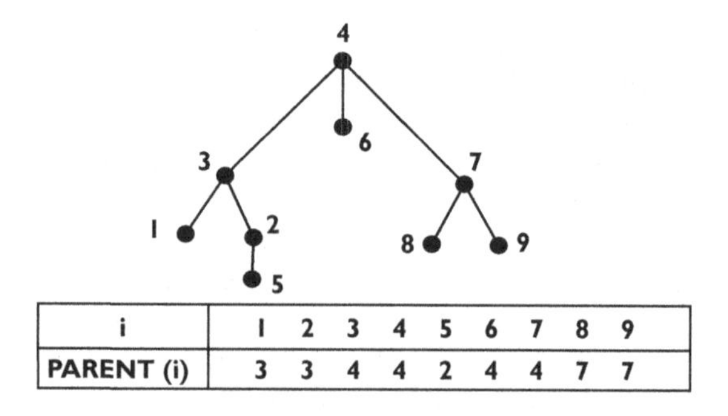

i	1	2	3	4	5	6	7	8	9
PARENT (i)	3	3	4	4	2	4	4	7	7

FIGURE 4.23 A root of a tree and its parent representation.

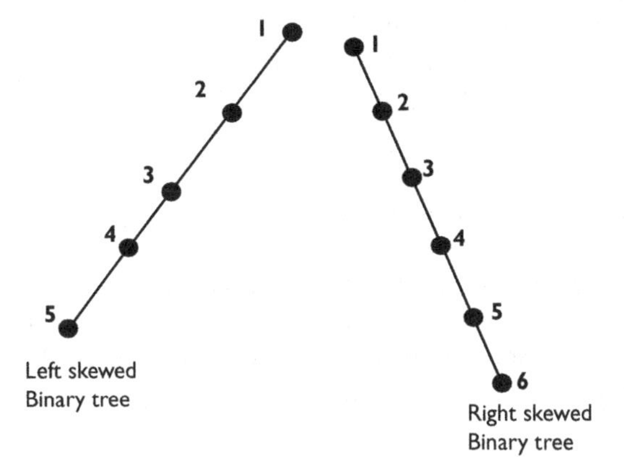

FIGURE 4.24 Skewed binary trees.

node are usually called the *left child* and the *right child*. A binary tree in which no vertex has a left child is said to be right-skewed. We define a left-skewed binary tree similarly. Figure 4.24 shows left-skewed and right-skewed binary tree. Consider the binary tree shown in Fig. 4.25. At level 0, the root alone is there. At level 1, there are two vertices. At level 2 there are four vertices, and at level 3 there are eight vertices. This observation leads to the following theorem.

Theorem 4.1

1. In a binary tree there are at the most two vertices at level i.
2. The maximum number of vertices in a tree of height k is 2^{k-1}.

Proof The proof of condition 1 (in Theorem 4.1) is on induction on the level number and the result is true for $i = 0$. If the result is true for a level i, there are at most 2^i vertices at level i. Each of those vertices can produce at the most two children for level $i + 1$. So, the maximum number of vertices at level $i + 1$ is $2^i + 1$. Hence the result of condition 1 follows by induction on i. The proof of condition 2 is based

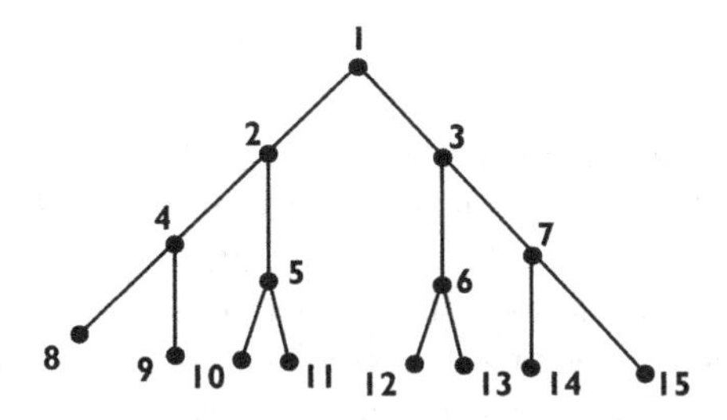

FIGURE 4.25 A full binary tree.

on the following numerical result:

$$2^0 + 2^1 + 2^2 + \cdots + 2^{k-1} = 2^k - 1 \tag{4.2}$$

A tree of height k with all the $2^k - 1$ nodes is called a *full binary tree*. In a full binary tree nodes can be numbered serially as follows: The root is assigned the serial number 1. The nodes at level 1 are numbered as 2 and 3 from left to right. After assignment of serial numbers for all the nodes at level -1, the consecutive numbers are assigned for all the nodes at level i from left to right. In Fig. 4.25, a full binary tree of height 3 and the serial numbering of its vertices are given. The following theorem can be easily verified for any full binary tree.

Theorem 4.2

In a full binary tree that is serially numbered

1. The left child of node i is $2i$.
2. The right child of node i is $2i + 1$.
3. The parent of i is $[i/2]$.

In a full binary tree, if some highest-numbered vertices are removed, it is called a *complete binary tree*. Figure 4.26a shows a complete binary tree; Fig. 4.26b, a full binary tree; and Fig. 4.26c, an incomplete binary tree. In a complete binary tree the numbering is continuous and coincides with the numbering of the full binary tree of the same height. Theorem 4.2. holds for any complete binary tree. Consider a binary tree in which every pendant vertex is assigned a positive weight. Let v and w denote the level number and weight of a pendant vertex, respectively. Consider the sum v_1, v_2, where the sum is taken over all pendant vertices. This sum is called the *weighted pathlength*. For example, consider the binary tree shown in Fig. 4.27. It has six pendant vertices: d,q,g,p,r,f. They are given weights 5,10,7,3,4,8, respectively. The weights and the level numbers are shown in Table 4.8. Given n pendant vertices and their weights, construction of a binary tree that minimizes the weighted pathlength is an interesting problem. It has applications in decision tree and optimal code construction problems. Some binary trees and their weighted pathlengths are shown in Fig. 4.28.

4.8.2 Spanning Trees

Let $G = (V, E)$ be a connected simple graph. A cycle-free, connected subgraph $T = (V)$, with all the vertices of G, is called a *spanning tree*. A graph G and some of its spanning trees are shown in Figs. 4.29a–4.29d.

Consider six cities, shown in Fig. 4.30a. The distance between the cities are represented by the weight of the edges. If a communication network has to be installed among the cities, it is sufficient if we establish the link along the edges of the spanning tree shown in Fig. 4.30b. The cost of installation of the communication

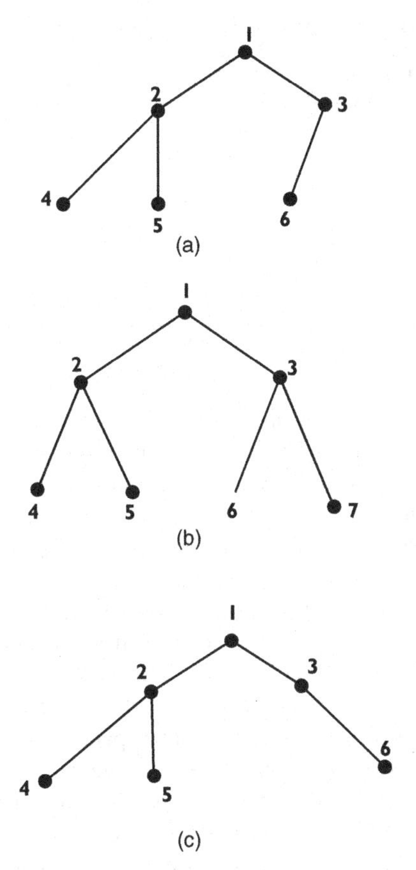

FIGURE 4.26 (a) A complete binary tree; (b) a full binary tree; (c) an incomplete binary tree.

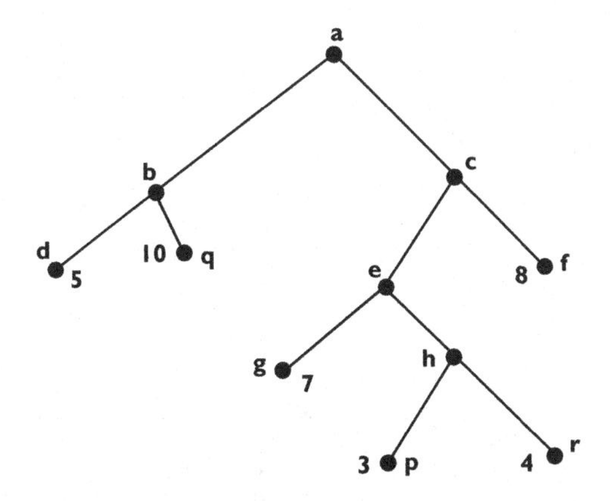

FIGURE 4.27 A tree with weighted pathlength = 95.

TABLE 4.8 Binary Tree Weights and Level Numbers

Pendant Vertex	Weight	Level	$l_i w_i$
d	5	2	10
q	10	2	20
g	7	3	21
p	3	4	12
r	4	4	16
f	8	2	16
		$\sum l_i w_i$	95

wires between two cities is proportional to the cost of the distance between them and, hence, the cost of the edge in the graph. Thus the total cost of installation is proportional to the sum of the weights of the edges. So we must find the spanning tree that has its sum of the weights of the edges' minimum. We define the weight of the spanning tree as the sum of the weights of its edges. Given an undirected graph, finding the minimum-weight spanning tree is a very interesting problem. The minimum-weight spanning tree is also called the *minimum-cost spanning tree* or *shortest spanning tree*. Every connected graph has a spanning tree of G. This tree, T, has $n - l$ edges. These $n - 1$ edges of the tree are called the *branches*. The nonfree edges are called the *chords*. If G is a disconnected graph (Fig. 4.31), it has more than one component. In this case, we can find one spanning subtree for each component, and the collection of such spanning subtrees is called the *spanning forest of G*. Figure 4.32 shows a spanning forest of the disconnected graph given in Fig. 4.31.

Let G be a graph with n vertices, m edges, and c components. The spanning forest of G contains $n - c$ branches, and there are $m - n + c$ chords. We define the rank of

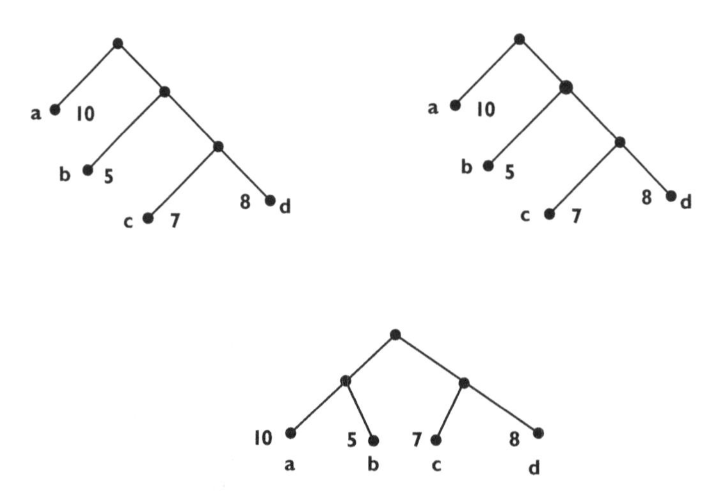

FIGURE 4.28 Some binary trees and their weighted pathlengths.

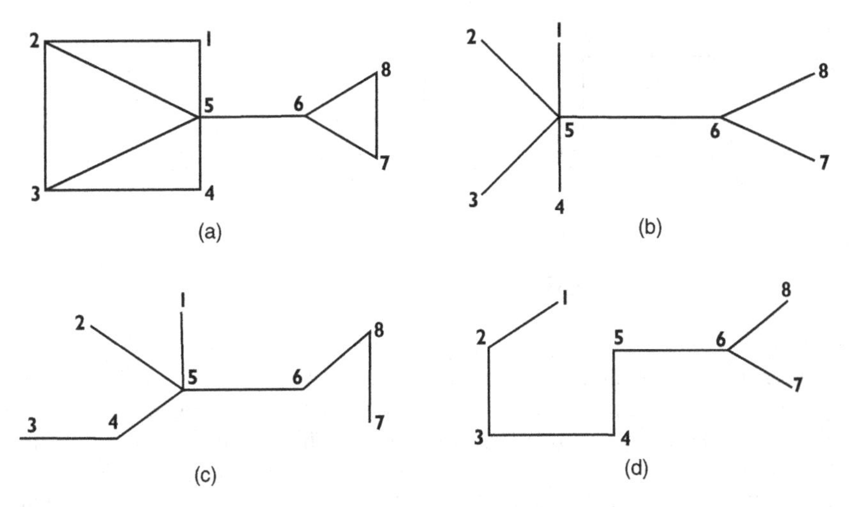

FIGURE 4.29 (a) A graph G; (b) spanning tree T; (c) spanning tree T_i; (d) spanning tree T_3.

the graph as the number of branches and the nullity of the graph as the number of chords. For the graph shown in Fig. 4.31a, $n = 15$, $m = 18$, $c = 3$, and

$$\text{Rank} = n - c = 15 - 3 = 12 \tag{4.3}$$

$$\text{Nullity} = m - n + c = 18 - 15 + 3 = 6. \tag{4.4}$$

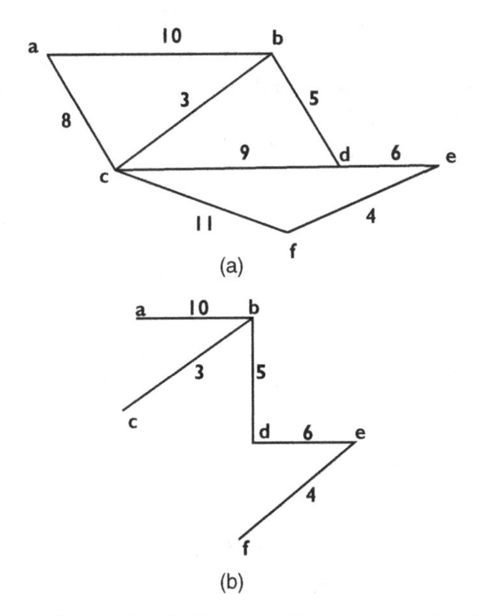

FIGURE 4.30 (a) Cities and their distances; (b) communication links for six cities.

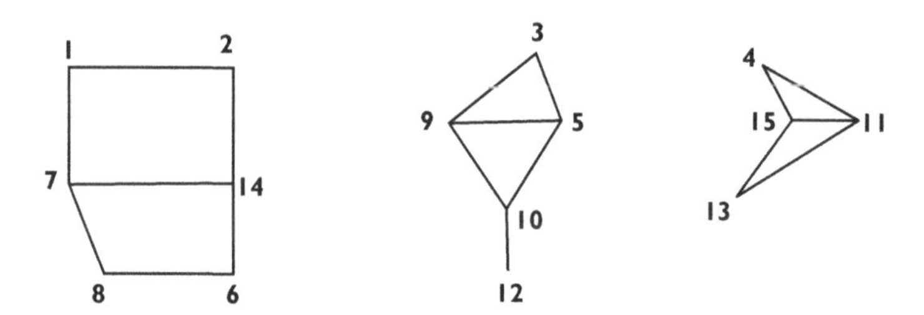

FIGURE 4.31 A disconnected graph G.

Notice that rank + nullity = number of edges. Also observe that for a connected graph, $n-1$ is the rank and $m-n+l$ is the nullity. Moreover, let us consider a connected graph along with a spanning tree T with "e" being a chord. If we include e in T, it will create a cycle. Here $b_1, b_2, \ldots, b_t$ are branches and e alone is the chord. This cycle is called a *fundamental cycle*. There are $m-n+1$ fundamental cycles. It is also interesting to count the number of spanning trees of a graph. Let G be a graph and T be its spanning tree. Let e be a chord and let $e(b_1, b_2, \ldots, b_t)$ be its fundamental cycle. By including e with T and removing from T any b_i, we obtain a different spanning tree [denoted as $T_i(e)$]. For every chord e, we can find these spanning trees $T_i(e), i = 1, 2, \ldots,$. We observe that any spanning tree can be obtained by repeating this operation on the given spanning tree T. A graph G and all its spanning trees are given in Figs. 4.33a–4.33c. The operation, including a chord to the spanning tree and deleting a branch from its fundamental circuit, is called *cycle interchange*. It is noted that, given any two spanning trees T and T_2 of the same graph G, the value of one spanning tree can be obtained from those of the other one by a finite number of cycle interchanges. The number of cycle interchanges needed to obtain one spanning tree value from those the other of one is equal to the number of edges in one spanning tree that are not in the other. We define $d(T_1, T_2)$ = number of cycle interchanges needed to get T_2 from T_1 = number of edges in T_1 that are not in T_2.

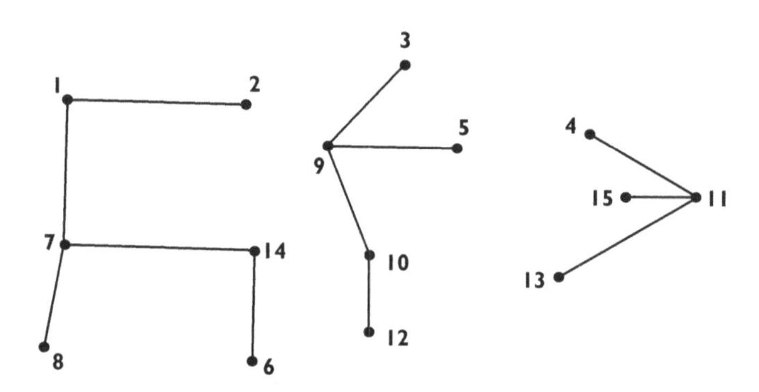

FIGURE 4.32 A spanning forest of G.

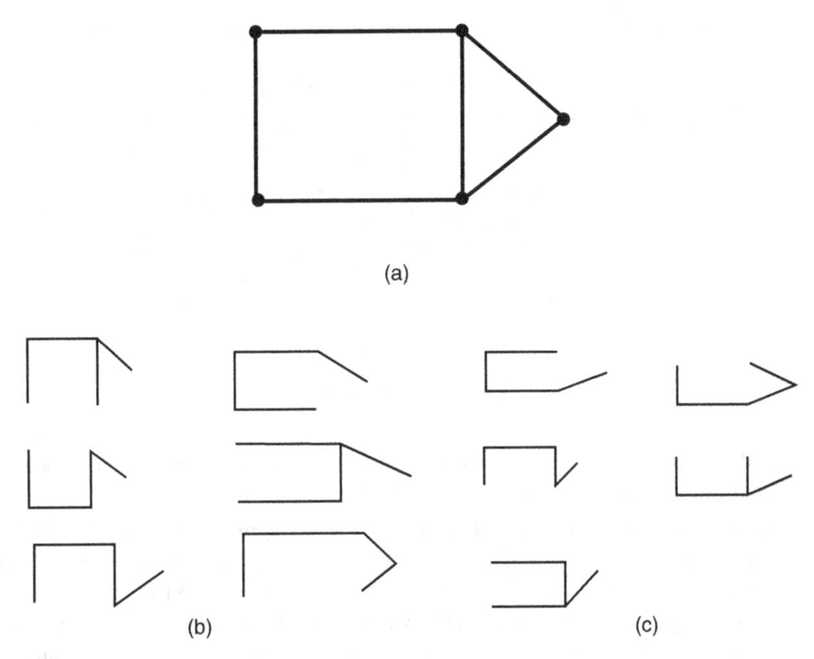

FIGURE 4.33 (a) A graph G; (b) some spanning trees of G; (c) some more spanning trees of G.

We can also observe that $d(T_1, T_2)$ satisfies the following properties for a metric:

1. $d(J,T_2) > Q$ and $d(T_x,T_2) = 0$ if and only if $T = T_2$.
2. $d(T_u T_2) = d(T_2,T_0)$.
3. $d(T_u T_2) < d(T,T_3) + d(T_3 + T_2)$ for any spanning tree T_3 other than T_x or T_2.

Also, $d(T, T_2)$ cannot exceed the rank or nullity of the graph. Let G be a connected graph and T be the collection of all the spanning trees of G. A graph can be formulate dusing J as a vertex set. Let the graph be denoted by $G_T = (T, E_T)$, where two vertices T_1 and T_2 are joined by an edge and only if $d(T_1, T_2) = 11$. The graph G_T is called the *tree graph* of G.

4.9 GRAPH TRAVERSAL

In order to visit all the vertices of a graph, there must be some systematic way to ensure that no vertex is left unvisited. There are two methods for traversing the vertices: (1) Breadth-First Search (BFS) and (2) Depth-First Search (DFS).

Let v be the starting vertex for traversing the graph G. The BFS starts by visiting the vertex v. After visiting v, all its adjacent vertices are visited in some defined order. Let the adjacent vertices of v, be $v, v_2, \ldots, v_d$. After visiting these vertices, we must

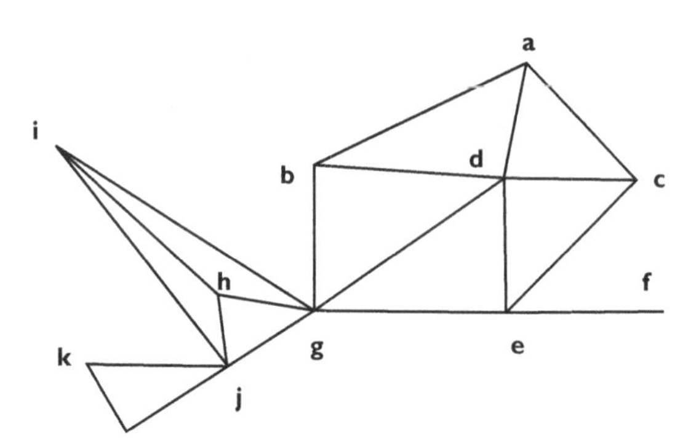

FIGURE 4.34 A graph.

visit all the unvisited vertices that are adjacent to these vertices. The procedure is repeated until all the vertices of the graph are visited.

In the DFS technique, we start from v. Then we visit an adjacent vertex v_1 of v. Now we visit an unvisited vertex adjacent to v_1. At some stage, if u has no unvisited vertex adjacent to it, we backtrack and come to V_{i-1} and visit an unvisited vertex adjacent to it. The procedure is repeated until we backtrack to v and visit all its adjacent vertices. The traversal forms a spanning tree for G. A graph G, its BFS spanning tree, and its DFS spanning trees are shown in Figs. 4.34 and 4.35. In the BFS spanning tree the dashed lines represent the chords and the solid lines, the branches. In the BFS spanning tree note that every chord joins vertices either at the same level or at consecutive levels. This property can be effectively used to solve several problems in graph theory.

4.10 CONNECTIVITY

If $G = (V,E)$ is a graph and $v \in V$, $G - v$ represents the graph induced by $V - v$. This is illustrated in Fig. 4.36. Let $G = (V, E)$ be a graph with k components, $a \in V$. If $G - a$ has more than k components, a is called an *articulation point*. In a connected graph a vertex whose removal disconnects the graph is also called an *articulation point*. In the graph shown in Fig. 4.36a, vertices 4 and 7 are the two articulation points. A connected graph that has no articulation point is called a *biconnected graph*. Any cycle is biconnected. Let $G = (V,E)$ be a graph and u, v be nonadjacent vertices of the same component of G. A subset S of V is called a *u–v separator* if the removal of S from G results in u and v lying on different components. In other words, S is a *u–v separator* if there is no path from u to v without passing through the vertices of S. A u–v separator that does not contain any other *u–v separator* is called a *minimal u–v separator*. A u–v separator having a minimum number of vertices is called a *minimum u–v separator*.

```
Let S1 = {1 ,2 ,3 ,4} S2 = {1 ,2 ,4} S3 = {2,4}
    S4 = {2 ,4 ,8} S5 = {8} S6 = {5}
```

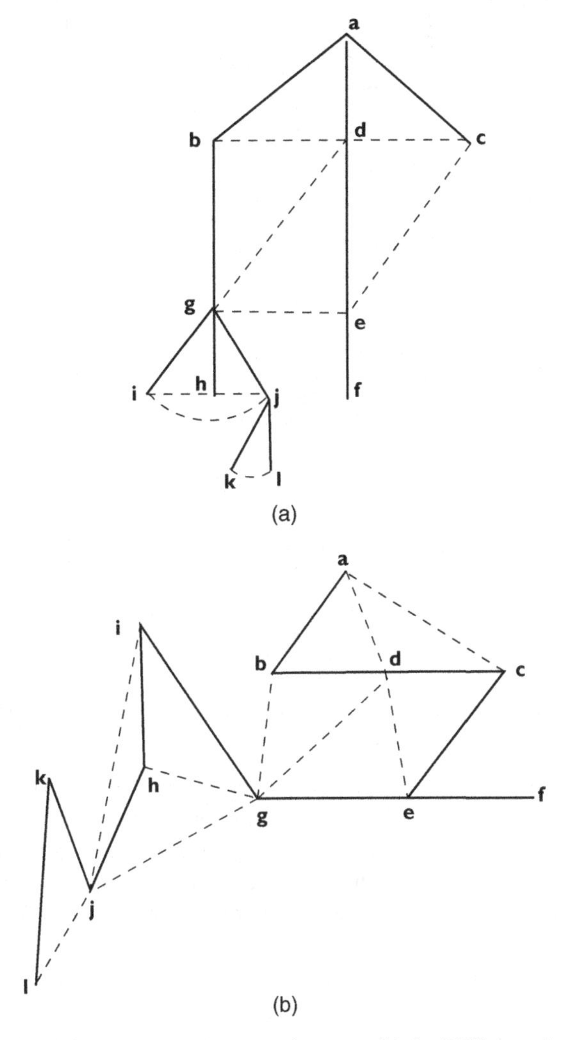

FIGURE 4.35 (a) The BFS spanning tree; (b) the DFS spanning tree.

Each term $S_1, S_2, S_3, S_4, S_5, S_6$ is a 7–6 separator in the graph G of Fig. 4.37. However, S_3 is a minimal 7–6 separator. Similarly, S_5 and S_6 are also minimal 7–6 separators. $S_5 = 8$ and $S_6 = 5$ are two minimum 7–6 separators.

Let $G = (V, E)$ be a connected graph and let S be a subset of V. If $G - S$ is disconnected, S is called a *separator of* G. A separator S is called a *minimal separator* if, whenever S' is a subset of S, S' is not a separator. A minimal separator of minimum cardinality is called a *minimum separator*. The cardinality of a minimum separator is defined as the vertex connectivity of the graph. In other words, the vertex connectivity of a connected graph is the minimum number of vertices to be removed from the graph, in order to make the remaining graph disconnected. Note that if a

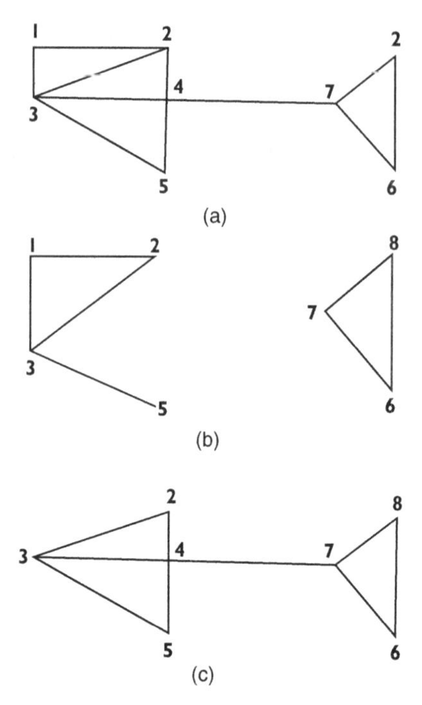

FIGURE 4.36 (a) Graph G; (b) graph $G - 4$; (c) graph $G - 1$.

graph has an articulation point, the vertex connectivity of the graph is 1. A graph with vertex connectivity 1 is called a *separable graph.* A nonseparable graph is also called a *biconnected graph.* For a biconnected graph the vertex connectivity is at least 2. A connected graph with vertex connectivity of at least 3 is called a triconnected graph. In other words, a triconnected graph is a connected graph in which the removal of

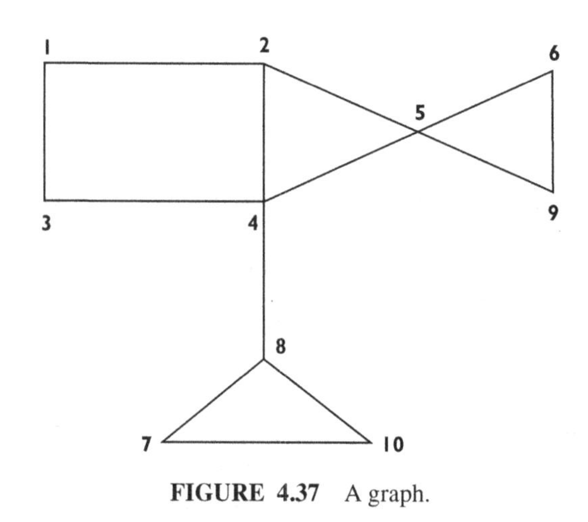

FIGURE 4.37 A graph.

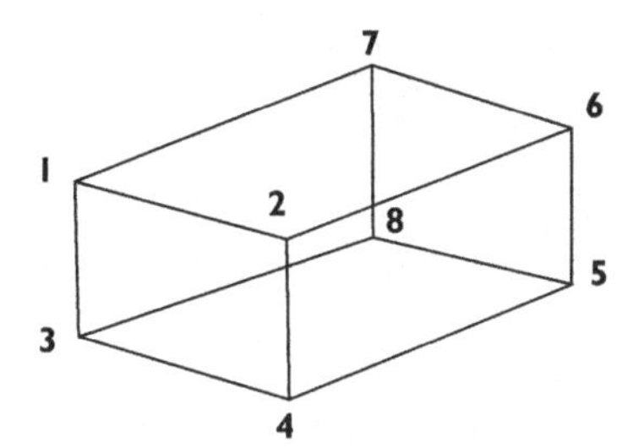

FIGURE 4.38 A triconnected graph.

any one or two vertices does not make it disconnected. Observe that any triconnected graph is biconnected and a biconnected graph is connected. A connected graph with vertex connectivity of at least k is said to be k-connected (where k is a positive integer). The following results immediately ensure:

1. A connected graph is biconnected if and only if for any two vertices u and v there are at least two edge-disjoint paths from u to v.
2. A connected graph is triconnected if and only if for two vertices u and v there are at least three edge-disjoint paths from u to v.
3. A connected graph is k-connected if and only if for any two vertices u and v there are at least k edge-disjoint paths from u to v.

The graphs shown in Fig. 4.36a and 4.37 are connected, but not biconnected. Any cycle C_n is biconnected. Figure 4.38 shows a triconnected graph. If d denotes the minimum degree of graph G, vertex connectivity of G is at most d.

We have already seen that a disconnected graph has more than one connected component (a connected component is a maximal connected subgraph). Similarly, we are defining the biconnected component as a maximal biconnected subgraph of the graph. If the graph itself is biconnected, there is only one biconnected component. Otherwise, the biconnected components can be found. Figure 4.39b shows the biconnected components of the graph shown in Fig. 4.39a. The biconnected components are also called *blocks*. A graph G in which every block is a complete subgraph is called a *block graph*. Figure 4.40a shows a block graph. The biconnected components of a graph G can be found by the following operation:

1. If G has no articulation point, G itself is the only biconnected component.
2. For every articulation point as V, let $V_1, V_2, V_3, \ldots, V_s$ denote the sets of vertices in the different components of $G - a$.
3. Let $G =$ graph induced by V, U a $(i = 1, 2, 3, \ldots, s)$.

These definitions and procedures can be generalized for connected components $(k > 2)$. A block graph can be represented by its *block-cut* vertex tree representation (*BC-tree*). For example, consider the block graph shown in Fig. 4.40a. Its blocks

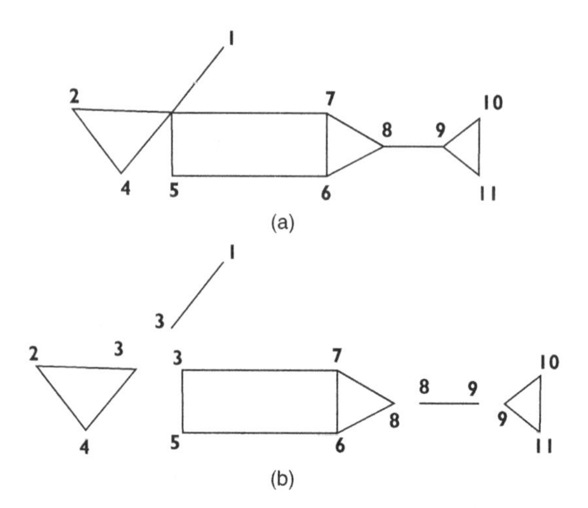

FIGURE 4.39 (a) A graph G that is connected but not biconnected; (b) biconnected components of G.

are $B_1 = 1{,}2$; $B_2 = 2{,}3{,}4{,}5$; $B_3 - 5{,}6{,}7$; $B_4 = 5{,}9$; $B_5 = 5{,}8$; $B_6 = 8{,}10$; $B_7 = 10{,}11$; $B_8 = 10{,}12{,}13$; $B_9 = 4{,}14$. The cut vertices are 2,4,5,8,10. The vertex set of the BC-tree consists of the blocks and the cut vertices. A block and a cut vertex are joined by an edge only when the vertex is contained in the block. The BC-tree of the block graph depicted in Fig. 4.40a is given in Fig. 4.40b.

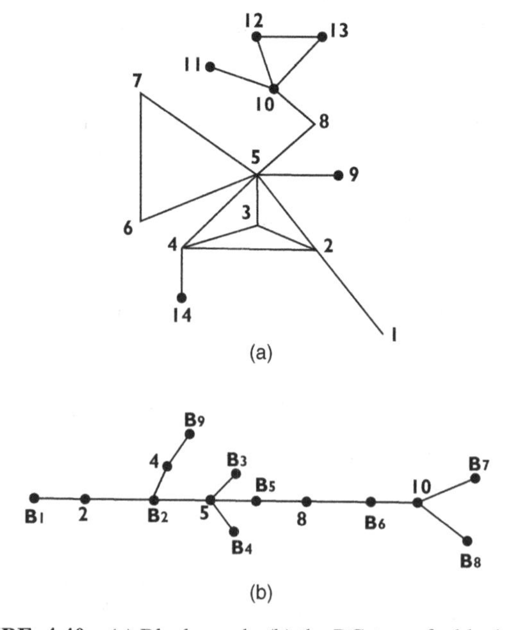

FIGURE 4.40 (a) Block graph; (b) the BC-tree of a block graph.

4.11 PLANAR GRAPHS

In this section, we define the graph G as a pair of two sets V and E, where V is any (nonempty) set and V is a subset of $V \times V$. Any graph can be schematically represented in more than one way. Figure 4.41 is a representation of the graph $G = (V, E)$, where

```
V= {a, b, c, d, e, f, g, h}
E = {{a, b), (b, c), (c, d),
     (d, a), (e, f (/, g),
     (g, h), (h, e), (a, e),
     (b, f), (c, g), (d, h)}
```

It is preferable to draw a graph on the two-dimensional (2D) plane so that no two edges intersect each other. It is evident that we cannot draw all the graphs in the plane in such a way that no two edges intersect. A graph is said to be planar if it can be drawn in a 2D plane without its edges intersecting. Figure 4.42 shows some simple examples of planar graphs. A graph $G = (V, E)$ is said to be *bipartite* if V can be partitioned into V_1 and V_2, such that every edge of G joins a vertex of V_i and a

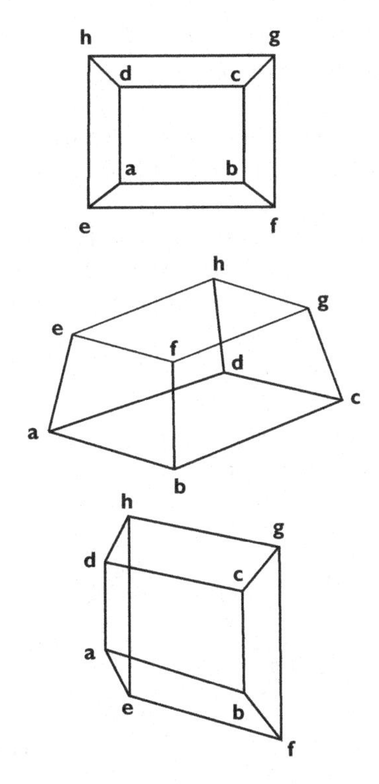

FIGURE 4.41 Same graph, different schematic representation.

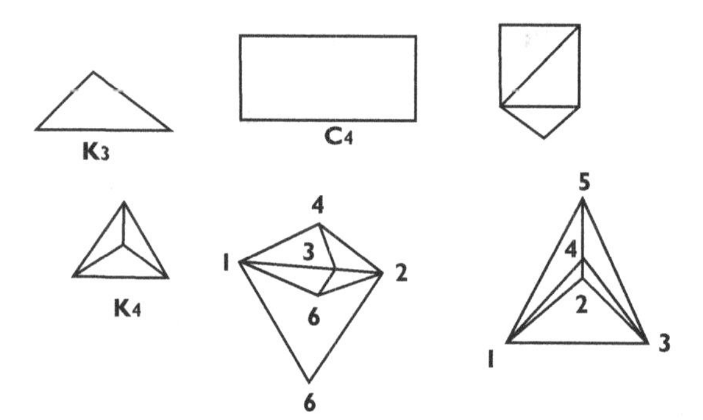

FIGURE 4.42 Some planar graphs.

vertex of V_2. Note that a bipartite graph contains no cycle of odd length. A complete bipartite graph is a bipartite graph in which every vertex of V is adjacent to all the vertices of V_2. A complete bipartite graph with m vertices in V_1 and n vertices in V_2 is denoted by K_{mn}. Note that K_{mn} has $m + n$ vertices and mn edges. The Polish mathematician Kazimierz Kuwatowski (1896–1980) proved that the following two graphs (see also Fig. 4.43) are nonplanar:

1. The complete graph with five vertices is given by K_5.
2. The complete bipartite graph $K_{3,3}$

Observe that both Kuratowski graphs are regular and nonplanar. Also note that they are minimal nonplanar graphs with respect to the number of vertices, as well as the number of edges.

Consider the planar graph G drawn in a plane (without its edges intersecting). The graph divides the plane into several regions. Figure 4.44 shows a planar graph and the region R_i, R_2, ..., R_e. It is easy to observe the following results:

1. Any simple planar graph can be embedded in a plane such that every edge is drawn as a straight-line segment.
   a. A planar graph can be embedded in a plane such that any specified region can be made the infinite region.

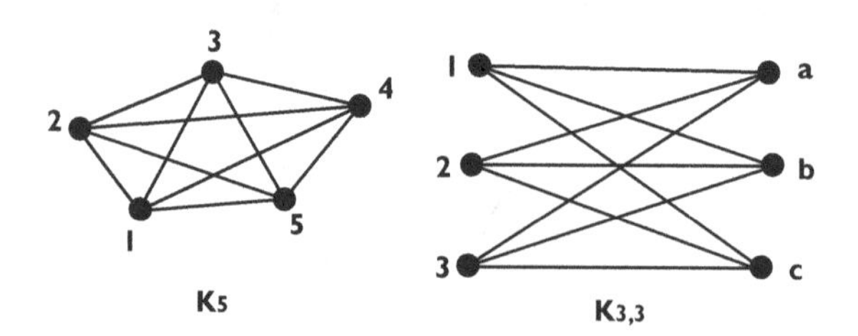

FIGURE 4.43 Kuratowski's graphs.

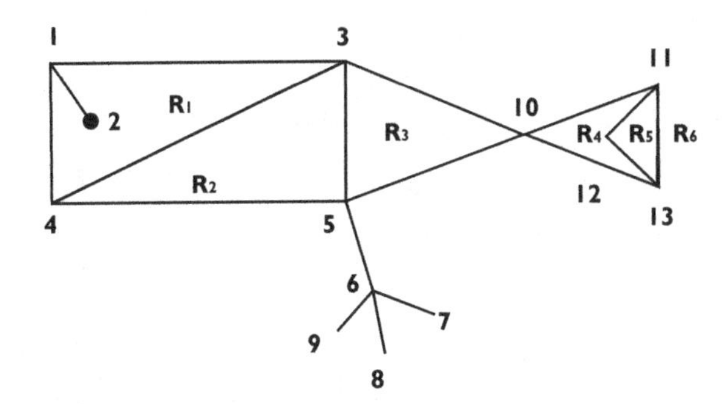

FIGURE 4.44 A planar graph with six regions.

b. Any planar graph can be embedded in a sphere, and any graph that can be embedded in a sphere is a planar graph.

2. If G is a connected planar graph with n vertices and m edges, it then would generate $m - n + 2$ regions. For example, in Fig. 4.44, $n = 13$, $m = 17$, and the number of regions is $m - n + 2 - 17 - 13 + 2 = 6$.

Also, we can prove that, for a planar graph, $m = 3n - 6$. This is a necessary but not sufficient, condition. Kuratowski's second graph is a nonplanar graph, satisfying this condition. A necessary and sufficient condition for a graph G to be planar is that G does not contain either of Kuratowski's two graphs or any other graph homeomorphic to either of them. An *outer planar graph* is a planar graph that has a planar embedding in which all the vertices of the graph lie on the exterior region (infinite region). A *maximal outer planar* (MOP) graph is an outer planar graph in which the addition of an edge between any pair of nonadjacent vertices will resultina non-outer-planar graph. A MOP graph is Hamiltonian.

Several applications of the concepts discussed in this section exist in sensor networks. Some examples of these include the location of holes in sensor network coverage and optimum deployment of sensors with the lowest number of nodes.

4.12 COLORING AND INDEPENDENCE

By *coloring* a graph, we mean assigning colors to each vertex of the graph. Some authors have studied the problem of coloring the edges of the graph. They call it *edge coloring*. Coloring the vertices is said to be vertex coloring. We restrict our discussion to vertex coloring and so, using the term *coloring*, we mean only vertex coloring. A "perfect" coloring is to assign colors to each of the vertices, such that any two adjacent vertices get different colors.

Figure 4.45 shows a graph G and three different perfect colorings for G. The main interest here is to color the graph using a minimum number of colors. Such a coloring is called a *perfect minimum coloring*, and the number of colors used for minimum coloring is called the *chromatic number*.

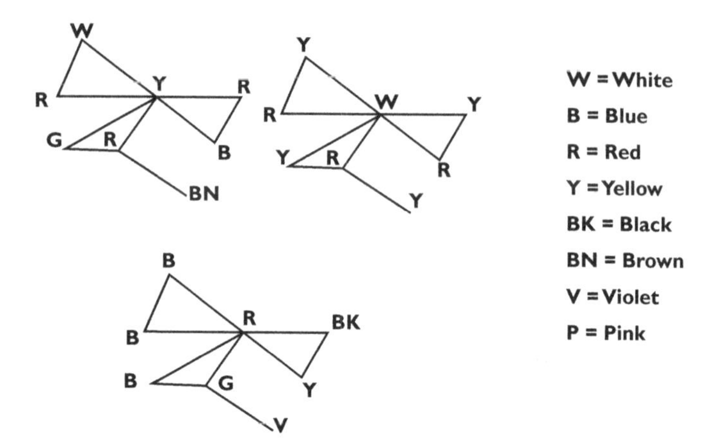

FIGURE 4.45 A graph with three perfect colorings.

The chromatic numbers of some well-known graphs are listed in the following table:

Sensor Number	Graph	Chromatic Number
1	K_n	n
2	Tree	2
3	A graph with no edge	1
4	A cycle of even length	2
5	A cycle of odd length	3
6	Bipartite graph	2

A complete subgraph is called a *clique*. A clique C with r vertices, is called an *r-clique*. A clique C is said to be maximal if there exists no other clique that properly contains C. The maximal clique of the greatest cardinality is called the *maximum clique*. The number of vertices in the maximum clique is called the *clique number*. In order to color an r clique, we need r colors. So, chromatic number = clique number.

4.13 CLIQUE COVERING

A clique cover of a graph $G = (V, E)$ is produced by partioning V into $V_1, V_2, \ldots, V_k$, such that each V, is a clique. In Table 4.9 we show two different clique coverings of the graph G shown in Fig. 4.46.

The size of a clique cover $V_i, V_2, \ldots, V$ refers to the number of partitioned sets k. The clique cover of minimum size is called the *minimum clique cover*. A set of vertices X of a graph $G = (V, E)$ is called an *independent set* if any two vertices of X are not adjacent in G. Independent sets are also called *stable sets*. A maximal

TABLE 4.9 Clique Coverings for graph G in Fig. 4.46

Clique Covering 1	Clique Covering 2
$V_1 = 1$	$V_1 = 1, 2$
$V_2 = 10$	$V_2 = 3, 4, 5$
$V_3 = 2, 8, 9$	$V_3 = 6, 7, 8$
$V_4 = 6, 7$	$V_4 = 9, 10$
$V_5 = 3, 4, 5$	

independent an V set is an independent set X, where $X \in v_{\text{is}}$ not an independent set for any vertex v not in X. The *maximal independent* set of largest cardinality is called the maximum independent set. For the graph shown in Fig. 4.46, some independent sets are shown below:

$$X_1 = \{1,4,8,10\}, \quad X_2 = \{1,10,6,3\}, \quad X_3 = \{8,4\}, \quad X_4 = \{1,7,10\}$$

Among these, X_1 and X_2 are maximum independent sets. All the vertices of an independent set can be colored with the same color. Each vertex of a clique needs to be colored with a distinct color.

4.14 INTERSECTION GRAPH

A number of data structures, graphs, and trees that have been discussed previously must also address a different need—integrating distributed data structures over sensor networks. However, for many deeply embedded systems we need a standard definition for a universal data structure that integrates various sensor network components efficiently into a single address space. That is the topic in the next section of the chapter.

Let F be a family of subsets. From the family F, we construct a graph $G = (V, E)$ as follows. The vertex set V has one–one (one-to-one) correspondence with F.

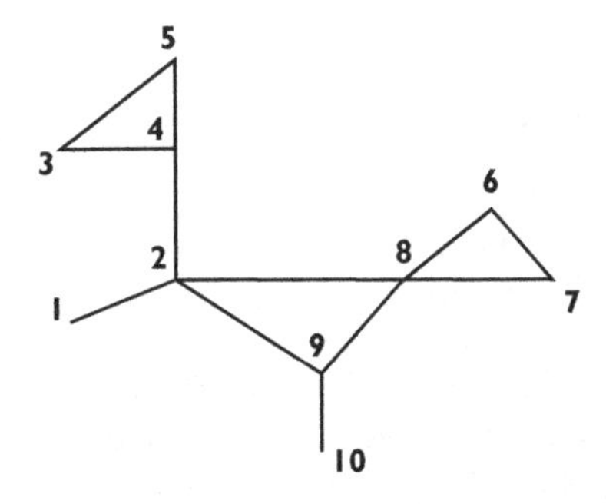

FIGURE 4.46 A graph with clique covering 1,2,3,4,5,6,7,8,9,10.

Two vertices of V are adjacent if and only if the corresponding subsets in F have a nonempty intersection. The graph G constructed P in this manner from is called the *intersection graph* of F. Let $F = P_1, P_2, P_3, P_4, P_5$, where $P = 1,2,3$; $P_2 = 2,3,4,5,6$; $P_3 = 5,6$; $P_4 = 6,7,8,9$; $P_5 = 4,10$. Let $(v_1, v_2, \ldots, v)$ be a cycle of a graph. Any edge joining two nonconsecutive vertices V_1 and V_2 is called a *chord*.

4.15 DEFINING DATA STRUCTURE OF SPANNING TREE PROTOCOLS

In the following paragraphs we define a new characterization of the data structure focusing on the programming aspects of sensor networks. To do this, we define flooding, replicated packet arrivals, and neighboring nodes all in the context of spanning tree data structures:

- *Flooding*. In its simplest definition, *floading* refers to a situation in which a networking device is overwhelmed with packets to such an extent that its processing capabilities are severely degraded. Most devices facing such a situation simply drop packets.
- *Neighboring Nodes*. The term *neighboring nodes* refers to sensor nodes in the same proximity as another node. Typically sensors collaborate with neighbroring nodes in processing tasks ranging from intermediate routing to computational tasks with their direct neighbors.

More details on these networking concepts are provided in the chapters on wireless sensor programming (e.g., Chapter 7).

4.15.1 Flooding

Owing to the broadcast nature of wireless sensor networks, messages can be easily send to a node's neighbors; then the neighboring nodes rebroadcast the packets to their neighbors within range. Flooding generates replicated packet arrivals to each node.

4.15.2 General Structure

The high-level structure uses a framework that has the capacity to send and receive messages. Each node has a unique (ID) that is used to initiate each broadcast. Because it is a broadcast, the destination is set as "FF," which allows the receiving node to rebroadcast the message untill all the nodes have received the messages from all other nodes. This creates numerous duplicates and chains in the network path. The type of data structure that is needed to store and optimize the minimal spanning tree would need to have a unique node ID, a parent ID, and two lists, one of which keep strack of new neighbors and another one that lists members of other, already existing, parents.

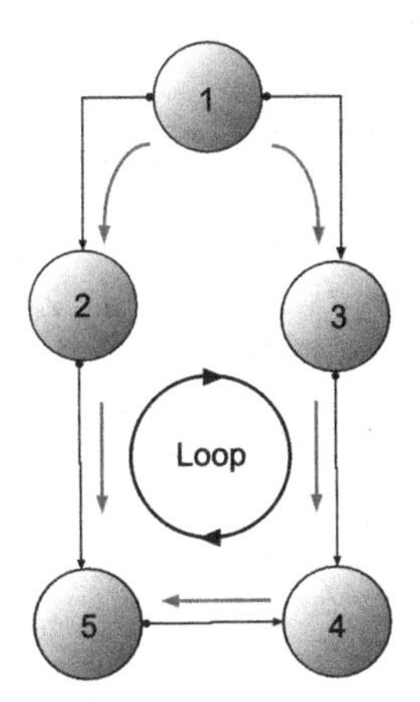

FIGURE 4.47 Flooding from source to destination.

If the totals of both the lists are counted, the sum will be equal to the number of edges in the network discovered. If all the edges equal maximum edges, then the program can stop.

4.15.3 Programming Problem

Formulating a flooding based protocol for tree construction can be a basis to prove that a general structure of message passing protocol in sensor networks can be used to discover a general topology for generic sensor networks. Proving the above assertion for a protocol programming, we would then eliminate cycles in the network and ultimately make it possible to achieve a pathway from source to destination (Fig. 4.47).

We need to specify the protocol for nodes $0 \cdots N$. Initially all nodes are reset and node 0 is initialized with an ID of 0 and root (r). Then the parent is set to root and it broadcasts messages to all nodes $0 \cdots N$. Once the messages are received by other nodes on the network that are currenly reset, it checks whether it has any parent ID; if not, it will reply ACK only to its new parent as a subscribed neighbor. If a node downstream has already been assigned, then it is a duplicate message and the node sends a NAK to its requesting parent. All initialized parents will receive and send ACK and will add the new neighbor to their lists. The protocol stops as soon as the number of neighbors is fully discovered. The spanning tree is shown in Fig. 4.48.

4.15.4 Pseudocode

- Messages types

M: request message

$P_i \rightarrow P_j$, valid parent: P_j is the parent of P_i

$P_i \rightarrow P_j$, invalid parent: P_i has already been selected as parent node

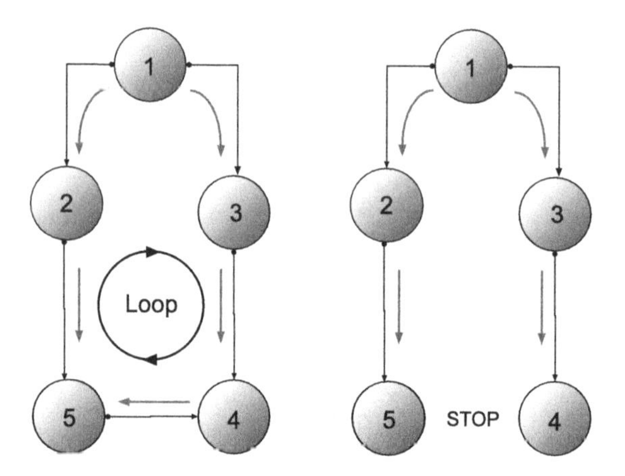

FIGURE 4.48 Spanning tree before (a) nonoptimal (b) optimal with no chains.

- Local data structure

  Neighbors: set of process IDs
  Children
  Node ID

- Event and actions

This pseudocode describes a simple way of embedding a basic concept of spanning tree data structure, (which was discussed in the data structures section.) Furthermore the structure explained in this example allows to explore a combination of simulation methodology and a code for a specific hardware driven topology of an actual sensor network using the idea of flooding.

```
{
INITIALLY:
        i: my own id;
        r: root node id;
        neighbors: Set of neighbors of i in the network;
        parent: NIL;
        children: empty set {0};
        others: 0;
START:
        if i = r and parent = NIL then
        Send M to all neighbors;
        parent = i;

        OnReceive (Message M)
```

```
switch (M->type)
{
case: NEIGHBORS
              if parent is NIL then
               Send <parent> to pj
              parent = pj
              else
               Send <reject> to pj
              break;
case: PARENT
                 Add Pj to children
                 if children UNION others contains
                       all neighbors except parent
                                      then terminate.
                 break;
case: REJECT:
                       Add Pj to others
                       if children UNION others contains
                             all neighbors except parent
                             then terminate.
                       break;
  }
}
```

4.15.5 Model Approach

We need to consider the case of synchronous versus asynchronous networks. In a synchronous nework the fooding message always starts from the root and foods to the destination nodes. The longest path on the network is also called the diameter D. In a asynchronous network the fooding messages can be initiated simultaneously from each and every node and reach neighbors that are within range. This is sometimes called *on-demand fooding*, where any node that needs to send data starts fooding to find its destination on the network.

4.15.6 Complexity

This allows us to compare the performance of both types of fooding protocol in terms of collision and maintenance of messages.

Message Overhead Message overhead complexity for both synchronous and asynchronous operations is

$$\text{Messages} = 2 \times E \times \text{number of message types} \tag{4.5}$$

where E is the number of edges in the network. This is similar for both types of network.

For the synchronous protocol model, time is measured as

$$\text{Messages} = O(D) \tag{4.6}$$

where D is the diameter. For a synchronous model

$$\text{Messages} = O(N) \tag{4.7}$$

where N is the total number of nodes in the network. In the worst-case scenario an async protocol may construct a *chain*. The synchronous network is more optimal as the diameter is always less than or equal to the total number of nodes.

PROBLEMS

4.1 What are some of the properties of data structures that enhance sensor programming in the context of distributed sensor networks?

4.2 Define a few efficient data structures for spanning-tree protocols.

4.3 Explain in your own words the concept of homeomorphism.

4.4 Illustrate using examples the concept of isomorphism.

4.5 What is the fundamental difference between a graph and a tree?

4.6 Consider the abstract data type *stack*. Specify clearly the interface operations for this data type. Make sure that you have accounted for all the exceptions that may occur.

4.7 Repeat Problem 4.1 for the data type tree. Choose any type of tree.

4.8 Consider two sensor nodes located adjacent to each other. We would like to build an abstract data type called *binary tree*, where the actual tree implementation may be distributed between the nodes. Give an interface specification for one such tree. Explain how you will implement the operations.

4.9 Consider a small network N_1 of wireless sensors. Draw the network as a graph G_1. Let n_1 be the node that acts as the manager node for N_1. Now consider several such networks $N_2, N_3, \ldots$, each represented by a graph $G_2, G_3, \ldots$ with their manager nodes $n_2, n_3, \ldots$. We want to manage the entire set of networks by organizing hierarchically all their managers. Show a simple management organization of the managers $n_1, n_2, \ldots$ that represents a hierarchy. In this hierarchy, how will you incorporate the other nodes in each network? Show the resulting data structure for a simple example.

REFERENCES

1. N. Chandrasekharan and S. Iyengar, NC algorithms for recognizing chordal graphs and k-trees, *IEEE Trans. Comput.* **37**:10 (1988).
2. S. S. Iyengar and R. Brooks, eds., *Distributed Sensor Networks,* CRC Press, (1995).
3. P. N. Klein, *Efficient Parallel Algorithms for Planar, Chordal and Interval Graphs,* Ph.D. Thesis, MIT, Cambridge, MA, 1988.
4. P. N. Klein, Efficient parallel algorithms for chordal graphs, *Proc. 29th IEEE Symp. Foundation of Computer Section, Proc. 29th* 1988, pp. 150–161.
5. S. Sastry and S. S. Iyengar, *Distributed Sensor Networks,* CRC Press, 2005.
6. A. S. Tanenbaum, *Computer Networks,* CRC Press, 2002.
7. C. Xavier and S. S. Iyengar, *Introduction to Parallel Algorithms,* Wiley, 1998.

5 Tiny Operating System (TinyOS)

> Simple things should be simple, complex things should be possible.
>
> —Alan Kay

An *operating system* (OS) can be defined as the master program existing between the computer hardware and the user. Its main purpose is the management and coordination of various system resources, which may include hardware resources such as hard disk, memory, peripheral device management, or nonhardware resources such as managing processor timeslots. It provides equal access to these resources through numerous system interfaces. Also, included in its role as a master program, an operating system can act as a host on which other programs can be executed, providing systemwide services by which these programs can access shared resources. Typical services provided by operating systems include networking services, file I/O, data security, and provision of user interfaces. There are several types of operating system, each differing only in its intended environment for use. Some common examples of operating systems include real-time, embedded, single-user, and multiuser operating systems. In most wireless sensor networks, sensors implement a basic operating system containing the minimum functions necessary to execute their tasks. This minimalist approach is of great importance because of the highly resource-constrained nature of these sensing devices; operating system code and application code must reside in devices having much less than one megabyte (1 MB) of memory. Sensor-based operating systems typically provide two basic services: a hardware abstraction layer for accessing the various sensing devices and a network interface for communication. There are several sensor-based OSes in use today; some of these are TinyOS, Contiki, Mantis OS, BTnut, and Nano-RK. In the following sections we discuss TinyOS and its associated programming language (nesC) in detail [1].

TinyOS, as its name implies, can be described as a miniature framework designed for embedded systems that require very aggressive resource management due to the highly constrained nature of their resources such as power and available memory [3]. It implements the hardware abstraction layer and scheduler of a conventional operating system, allowing generic programs that may have no knowledge of the intricate details of the operations supported by the underlying hardware components (such as sensors) to use well-defined interfaces to interact with these components. Its C-language-like framework provides an interface to core system components,

Fundamentals of Sensor Network Programming: Applications and Technology, By S. S. Iyengar, N. Parameshwaran, V. V. Phoha, N. Balakrishnan, and C. D. Okoye

allowing a programmer to manage various services of the system. The two basic software components/abstractions that constitute TinyOS are described below.

5.1 COMPONENTS OF TinyOS

TinyOS provides software abstraction for hardware components such as its communication, routing, sensing, and storage subsystems. *Software components* in TinyOS refer to abstractions of specific services either provided by either another software component or a hardware component. A software component consists of any number of the following:

- Modules
- Configurations

5.1.1 Modules

Modules are the lowest form of abstraction provided by the TinyOS operating system. They implement program logic that directly addresses a software component and can also provide a particular set of services, thereby enabling the reuse of software components. Other software components can interact with a module through its defined interface, which specifies a set of operations/services that a particular module implementation provides [3].

5.1.2 Configurations

On the other hand, abstractions of multiple modules and other configurations grouped together form a newly abstracted component referred to as a *configuration*. A configuration can be visualized as a supercomponent consisting of several subcomponents to provide a single unified interface. Configurations wire a set of components defined in a component signature—this is the set of interfaces that a component uses and provides to another component—thereby allowing two or more components to communicate with each other [3].

5.2 AN INTRODUCTION TO NesC

In TinyOS the software component abstractions (modules and configurations) [3] are written in a dialect of the C language called *nesC*. NesC contains a subset of features of the standard C libraries and syntax with some extensions such as commands, and events added to accommodate its event-driven approach to programming rather than the traditional imperative style of many C-based languages. Due to the resource-constrained environments typical of most sensors, these additional features in nesC help improve the efficiency of programs written in the language. These new constructs are outlined in Figs. 5.1–5.3.

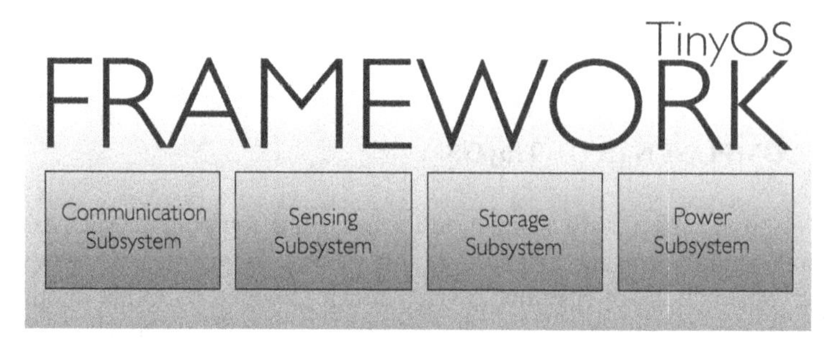

FIGURE 5.1 TinyOS system services.

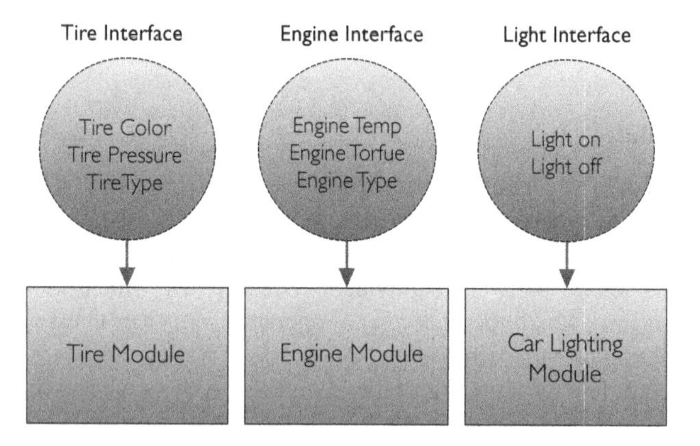

FIGURE 5.2 Three sample modules and their associated interfaces.

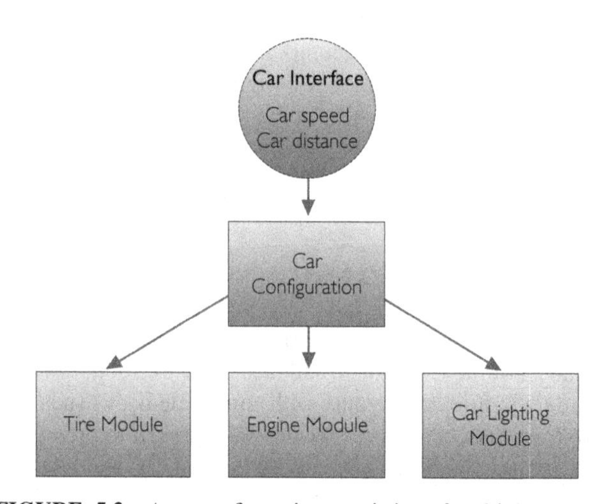

FIGURE 5.3 Acar configuration consisting of multiple modules.

5.2.1 Split-Phase Operations: Commands and Events

In addition to the standard C-function declaration, functions can be declared as commands, events or tasks.

```
command <return type> <command name> (<input parameters>)
command error_t  readSensor ()

event void <event name> (<returned messages and errors>)
event void readDone (error_t result, uint16_t value)
```

Commands are functions already implemented by the called software component in which its invocation and completion paths have been disassociated. In other words, most commands when called return immediately without completion, and if any errors or messages result at the end of its computation, a callback function is signaled. The notion of having a separate callback function signaled after the completion of a command is called *split-phase*. This is particularly useful since TinyOS contains no blocking operations; therefore any long-running operation can immediately return control to the calling program, and on completion a callback (signaling the event associated with the command) function is signaled.

A command and Its Associated Callback Function Events are triggered in response to some action; this may be after the execution of a command or in response to messages from the environment. For example, the reception of a packet triggers an event *receive*. Typically, events are not implemented by the called component but rather are functions whose implementation details are left to the calling software component. Simply put, events are methods that must be overridden by the caller. All callback functions signaled after the termination of a command are events, in which case their role would be to process the result of the terminated command.

Code Sample Illustrating the Split-Phase Concept In the blocking portion of the code presented above, the program does not increment the `SendCount` variable until the `Read` operation has completed and the conditional statement evaluated. This program's semantics are very different in the split-phase version of the code, in which the `Read` command is called but does not halt the program flow. When the reading operation is finalized, the event `ReadDone` is signaled with the value of the reading and a message indicating whether the reading is a success or failure.

5.2.2 Tasks

Finally, a task function represents the TinyOS notion of concurrency. It allows a function call to be deferred for a later time when the scheduler can process less important tasks. Multiple tasks can be queued up in the task queue, which is a FIFO (First-in/First-out) data structure. Unlike commands and events, a running task can be preempted by the system when an interrupt occurs.

```
// Blocking Non Split-Phase
if (Read () == SUCCESS)
{
        SendCount++;
}
//NonBlocking Split-Phase
//Start Phase
Sensor.Read ();

//Complete phase
event void ReadDone(error_t err, uint16 value)
{
        if (err == SUCCESS)
        {
                printf("Value read by sensor is \%d", value);
        }
}
```

Syntax of a Task Function

```
task <return type> <task name> (<input param>)
task void computation (void)
```

5.3 EVENT-DRIVEN PROGRAMMING

Event-driven programming (or *event-based programming*) refers to a programming paradigm in which program flow is determined by events [2, 3]. In wireless sensor networks, where conservation of computational and energy resources is paramount, event-based programming is very critical. In the absence of an event-driven approach, programs running on these sensing devices would have to continuously poll the system, checking for the occurence of certain events. This approach would be both wasteful of system resources and highly ineffective in larger sensor networks, causing program degradation. In TinyOS the event-driven programming approach allows for fine-grained control over power usage while maintaining the scheduling flexibility required in sensor networks due to their unpredictable nature. The following code snippet shows how events can be triggered and handled in a nesC program:

```
if (call AMSend. send (packet, sizeof (Data)))
{
                busy = TRUE;
}
event message_t* Receive.receive (message_t* msg, void* pay-
load, uint8_t len)
{               call Leds.led0Toggle;
                busy = FALSE
}
```

In this code, a node sends a data packet using its radio and the send interface while setting the busy flag to true. The data sent by node 1 triggers a corresponding receive operation on all nodes within the transmission radius of the first node without the need for continuous polling.

For a more complete description of TinyOS programming, the text by Levis and Gay see [3].

PROBLEMS

5.1 Briefly explain the components of TinyOS.

5.2 Distinguish between modules and configuration.

5.3 For the car example illustrated in Figs. 5.2 and 5.3, give nesC code to specify the "car" interface, "car" configuration, "tire" module, "engine" module, and "car lighting" module.

5.4 Extend the nesC code in problem 5.3 by adding "brake" and "engine" modules to the "car" configuration and by adding the "driver" configuration to the "car" interface.

5.5 Explain the *split-phase* concept with an example.

5.6 Explain the *task* function with an example.

5.7 What is event-driven programming, and why is it critical for sensor network programming?

5.8 Briefly discuss the following topics pertaining to nesC:

(a) Concurrency

(b) Synchronous code

(c) Asynchronous code

5.9 Specify a nesC interface for the following data types: stack, queue, singly linked list, and binary tree.

5.10 Consider a data stucture composed of a stack and queue, and assume that we want to build a datatype capable of manipulating the stack and queue individually. Propose a nesC interface for this data structure.

REFERENCES

1. J. Hill, R. Szewczyk, A. Woo, S. Hollar, D. Culler, and K. Pister, System architecture directions for networked sensors, In *In Architectural Support for Programming Languages and Operating Systems,* 2000, pp. 93–104.
2. R. Kumar and V. Garg, Modeling and Control Logical Discrete Event Systems, Kluwer Academic Press, 1995.
3. P. Levis and D. Gay, TinyOs Programming, Cambridge Univ. Press, 2009.

6 Programming in NesC

Low-level programming is good for the programmer's soul.

—John Carmack

6.1 NesC PROGRAMMING

Because of the highly resource-constrained nature and unpredictability associated with wireless sensor networks, traditional programming techniques cannot be used in the development of programs intended for use in these networks. Unlike conventional programming techniques in which most resources are available locally and remote resources could be accessed through persistent connections, sensor applications have to contend with several factors that may lead to degradation in performance of the sensor network, ranging from energy depletion, to limited or no connectivity, to physical damage of the sensor hardware in hostile environments. In TinyOS-based sensor networks a C-based language called *nesC* has been developed with several constructs built in to mitigate some of these undesirable properties of sensor networks. NesC shares a similar syntax with C but runs only on the TinyOS operating system. The language is simple so that it can be used for programming embedded systems, wireless sensor nodes, and similar applications, where efficiency in terms of speed and memory is of utmost importance. The nesC language has features borrowed from computer hardware such as the notion of components and subcomponents, each interacting with the other via clearly defined interfaces, and the concept of split-phase operation. We will present concepts in the nesC language using a sequence of simple examples where we will attempt to explain one chosen concept at a time.

6.2 A SIMPLE PROGRAM

A typical nesC program (henceforth referred to as a *component*) consists of an interface, a module, or several modules and a configuration file showing the connections between the various components of the modules. An *interface* consists of a set of clearly defined commands and event signatures that represent how interactions can

Fundamentals of Sensor Network Programming: Applications and Technology, By S. S. Iyengar, N. Parameshwaran, V. V. Phoha, N. Balakrishnan, and C. D. Okoye

occur between multiple modules [1]. An example showing how a simple *PrintMessage* interface is implemented is shown below.

PrintMessage.nc. This interface, called *PrintMessage*, consists of a command *send* and an associated event *send-Done*. Any component making use of this interface can call the command *send* with the appropriate parameters and expect the corresponding *sendDone* event to be signaled if there are no errors during execution. The actual implementation of the command *send* is specified in the PrintMessageC module shown below. By design, commands are implemented in the "called" components while the "caller" handles the implementation of the events.

```
interface PrintMessage
{
command error_t send (char msg[20], uint8_t len);
event void sendDone (char msg[20], error_t err);
}
```

PrintMessageC.nc. In terms of *code*, this is a version of the "HelloWorld" program for sensor networks. When the *send* command is called from another component, a message passed to the send command is printed on the screen using the built-in debug function. Note that this message can been seen only when run in a simulator since sensor nodes are devoid of any output except for some debug LEDs. On printing, the *sendDone* event is signaled to indicate completion of the command. This is another illustration of the callback function discussed in the preceding chapter. As stated before, events are handled by the calling program, in this case PrintClientC below, which "calls" the PrintMessageC component. The rationale for the implementation of events in the caller is because a module making use of an interface provided by another component is in a better position to deal with events generated due to the execution of a command. Thus it can be said that PrintClientC makes use of the interface PrintMessage and a particular implementation of PrintMessage is provided by the component PrintMessageC.

```
module PrintMessageC
{
}
implementation
{
       command error_t PrintMessage.send (char msg[20],
       uint8_t len)
       error_t err;
       if(strlen (msg) < 20 && len < 20)
       {
                dbg("STD", "Message_was_\%s\n" ,msg);
                err = SUCCESS;
                signal PrintMessage.sendDone (msg, err);
       }
```

```
            else
            {
            err = FAIL;
            signal PrintMessage.sendDone (msg, err);
            }
    }
```

PrintClientC. This is a client module that uses a set of interfaces to print a *Hello World* message. In this case, it uses two interfaces `Boot` and `PrintMessage` and thus provides implementation for all the events defined in these two interfaces. The nesC code for the `Boot` interface is shown below.

```
interface Boot {
event void booted ();
}
```

The *Boot* interface is synonymous with the main function in traditional C programs providing a point from which program execution starts, that is, by first initializing all components necessary for successful program execution.

```
module PrintClientC
{
        uses interface Boot;
        uses interface PrintMessage;
}
implementation
{
        event void Boot.booted ()
        {
                call PrintMessage.send ("Hello World", 11);
        }

        event void PrintMessage.sendDone (char msg[20], error_t err)
        {
                if (err == SUCCESS)
                {
                        dbg ("std", "SendDone signalled with no errors \n");
                }
                else
                {
                        dbg ("std", "SendDone signalled with an error \n");
                }
        }
}
```

So far we have defined an interface and implemented all our commands in one module and the events in another module. Although our program (consisting of two interfaces and two modules) is logically ready to run, there is one more issue that must be resolved before the program can execute successfully. The PrintClientC module uses two interfaces (Boot, PrintMessage); therefore, we must specify which components provide the particular implementation of these interfaces since any of these interfaces could be provided by multiple components. Which interface is implemented by which component is specified by *wiring* [1]. Each component contains a configuration file in which the wiring details are specified. An example of a typical wiring is shown below:

$$PrintClientC.PrintMessage \rightarrow PrintMessageC.PrintMessage.$$

PrintClientC. PrintMessage is said to have been wired to *PrintMessageC. PrintMessage*. It signifies the fact that the interface *PrintMessage* used in the application *PrintClientC* uses the implementation of this interface provided by the module *PrintMessageC*. Similarly, the wiring statement *PrintClientC.Boot* → *MainC.Boot* shows that the Boot interface used by PrintClientC is provided by Boot in MainC, where MainC is the system component. This wiring is indicated in a module called the *configuration module*, described below.

PrintClientAppC.nc. The configuration module first lists all the components whose names appear in this module and then gives the wiring connections (as we discussed above). Wiring is discussed in more detail in Section 6.2.5.

With this our application code is complete, and we are now ready to execute the program. By convention, the TinyOS system invokes a command somewhere, which results in the invocation of the event *booted* implemented in the module PrintClientC. Thus the system starts executing the event *booted*, which is where our program execution begins as far our program is concerned. Once the execution of *booted* is started, we call `PrintMessage.send`, which prints the message "Hello World." After printing, this command signals back to invoke the event PrintMessage.sendDone to indicate whether the command was executed correctly, and prints a message appropriate for the error reported by the command. We can now show the overall structure of our program (rather, component) using the component diagram in Fig 6.1.

```
configuration PrintClientAppC
{
}
implementation
{
        components MainC;
        components PrintMessageC;
        components PrintClientC;

        PrintClientC.Boot -> MainC;
        PrintClientC.PrintMessage -> PrintMessageC;
}
```

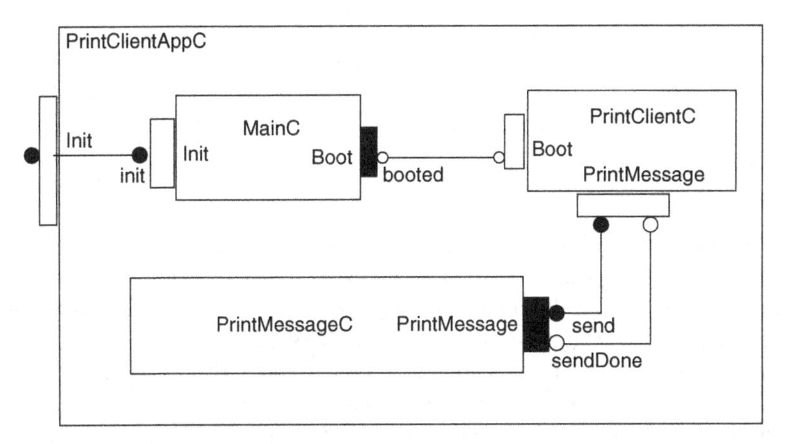

FIGURE 6.1 Component architecture for PrintClientAppC.

Although this program is a simple one, it illustrates several issues that one has to deal with while writing a nesC program, which can be summarized as follows:

Define an interface I.
Provide implementation for the commands from the interface in a module M_1.
Provide implementation for the event from the interface in a module M_2 that invokes the commands.
Provide a configuration module showing the wiring details.

It is now easy to modify our initial implementation and interface built for PrintMessage to perform more complex tasks. The interface PrintMessage-multi shown below adds a new sending command (send2) and its associated event send2-Done.

PrintMessage-multi.nc. Implementing this new interface requires the addition of one new command in our PrintMessage implementation in which the event send2-Done will be signaled. By now, the role of events should be more apparent in interface declarations. The notion of commands is fairly easy to understand. Commands are functions that a component C calls in order to achieve some result. Execution of a command might produce the desired result in addition to other exceptions. What to do with the result and the exceptions is not the command executor's job; rather, it is the caller C's responsibility, and what exactly C would like to do with the results and the exceptions should be defined in the component C. In the current example, we can imagine that the event send1-Done corresponds to the command send1, and the event send2-Done corresponds to the command send2.

```
interface PrintMessage-multi{
        command error_t send1(char msg[20]> uint8_t len);
        command error_t send2(char msg[20]> uint8_t len);
```

```
        event void send1-Done(char msg[20]> error_t err);
        event void send2-Done(char msg[20]> error_t err);
}
```

6.2.1 Tasks

As we have seen before, the completion of an operation, in the split-phase strategy, is signaled by invoking a corresponding event. Initially, it might seem harmless to signal an event from a different component at the end of the code that implements the operation (this is what we have done in PrintMessageC.nc above). Serious problems can sometimes arise with this approach. For example, if the code for a command C makes a direct call to its corresponding event, and the event in turn makes a call to the command C, then this might cause a potentially long loop, causing the system stack to overflow. We can avoid this by using a task, which is a deferred procedure call [1]. The command C, instead of signaling its corresponding event *e* immediately, can create a task that signals the invocation of the event *e*, and then post it on the system queue. The task will eventually be executed, thus potentially avoiding the creation of a long loop. PrintMEssageC below shows how to implement *sendDone* as a task.

```
module PrintMessageC {
provides interface PrintMessage;
}
implementation {
command error_t PrintMessage.send (char msg[20], uint8_t len)
{
error_t err;
if (strlen(msg)< 20 && len < 20)
   {
   dbg("std","Message_was_%s\n",msg);
   err = SUCCESS;
   post sendDoneTask(msg, err);
   return err;
   }
else {
err = FAIL;
post sendDoneTask (msg,err);
return err;
} // end commad

} // end implementation


task void sendDoneTask (char msg[20], err) {
signal PrintMessage.sendDone(msg, err);
}
```

PrintMessageC Implementation A task when posted goes into the queue and waits in the queue until it is dequeued and serviced by the system. Long-running tasks can be broken down into chunks and reposted on the queue. Tasks are defined to be non-preemptive with respect to one another. In other words, if a task is running, then any other tasks ready to run will have to wait until the currently running task is completed. So, if there were any shared variables, their values will be protected from any other tasks changing it. For this reason, it is often advisable that tasks be reasonably small so that they can run to completion quickly in order to allow the waiting tasks to start running.

6.2.2 Asynchronous Commands and Events

The commands in an interface can be classified into two types:

- Asynchronous
- Synchronous

The *async* keyword preceding a command name denotes a command that may run in asynchronous context. By default, all command declarations in TinyOS are synchronous [1]:

```
interface PrintMessage-multi<val_t>
{
        async command error_t send1(char msg[20], <val_t> len);
        async command error_t send2(char msg[20], <val_t> len);
        async event void send1-Done(char msg[20], error_t err);
        async event void send2-Done(char msg[20], error_t err);
}
```

This code redefines our earlier interface `PrintMessage-multi` making each command and event asynchronous. An async command or event runs preemptively in the sense that it can be preempted before completion. All async elements can only interact with other async components. For example, an asynchronous command can only call other asynchronous commands. When a need to signal an event arises (signaling is synchronous in nature), the asynchronous command can post a task (asynchronous in nature), which then signals the event. By default, all code written in nesC is synchronous, but the addition of the keyword *async* denotes it as being asynchronous. Asynchronous operations usually have short execution cycles. Some examples of asynchronous elements in TinyOS are interrupt handlers. In the following sections, we discuss the main drawback of having asynchronous elements and ways to mitigate it.

6.2.3 Preemption Problems

One issue with async functions is that when an async function is preempted by another function modifying the same shared variables, it can affect the underlying computation, producing erroneous results. Consider the following example.

```
int x = 1;
async command int add() {
x = x +1;
// interrupt at this line #1.
  if (x % 2 == 0) return TRUE;
  else return! FALSE;
}
```

When this function runs without interruption for the first time, it returns TRUE. Suppose that it is preempted at line 1 by itself; then the new call of the function will increment the value of x again, making it 3, and the new call will return FALSE. After this, when the previous call continues, it will find the value of x to be 3 (erroneous) and the function will return FALSE instead of TRUE. One traditional solution to this problem is to make sure that when variables such as those stated above are shared across multiple processes, they must be protected, and this is done in nesC by declaring "atomic" blocks of code.

6.2.4 Atomic Block

An atomic block of statements is a sequence of statements that are executed to completion without preemption (being interrupted) by the TinyOS system. When async functions are used, we need to be careful about variables shared among multiple components. In such situations, atomic blocks of statements are useful. On the other hand, excessive use of atomic blocks of statements can degrade the performance of the TinyOS system. Finally, care must be taken to ensure that cyclic dependences do not occur in atomic code since they lead to deadlocks in the system. Thus, programs using async commands must make use of atomic blocks only when shared variables are used. An example of an atomic statement is presented below.

```
int x = 1;
async command int add()
{
  atomic
  {
  x = x+1;
  if ((x % 2) == 0) return TRUE;
  else return FALSE;
  }
}
```

In this example, multiple invocation of the async command share the values of the variable x. However, for a given invocation, the atomic block is not preemptable, and thus we will clearly know what result this function is returning. Note that atomicity does not necessarily mean total absence of preemption.

6.2.5 Wiring

In our earlier example we explained how to perform wiring so that module commands and events can be appropriately called during execution. There are a few more special cases that still need discussion.

```
configuration CC2420TransmitC
{
  provides interface Init;
}
  implementation
{
  components Alarm, CC2420TransmitP;
  Init = Alarm;
  Init = CC2420TransmitP;
}
```

Fanout In Fig 6.2, we have shown four components, CC2420TransmitC, CC2420TransmitP, Init, Alarm, and TransmitApp corresponding to `configuration CC2420TransmitC` shown (using nesC code) above. The TransmitApp component makes use of the Init interface provided by CC2420TransmitC.

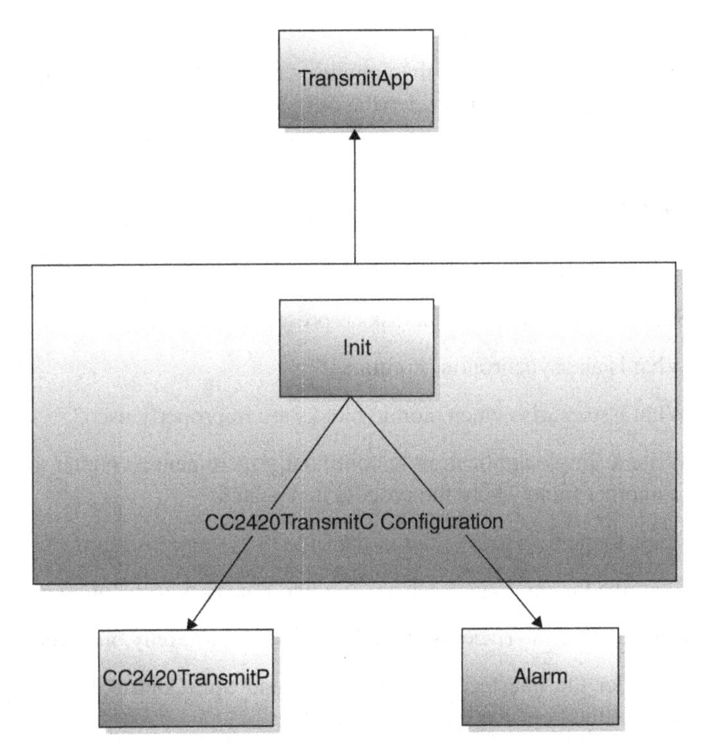

FIGURE 6.2 Flowchart illustrating the concept of fanout wiring.

This Init interface is wired to the interfaces provided by components Alarm and CC2420TransmitP. Therefore, when a command is called from TransmitApp to the Init interface, it fires the respective commands in both Alarm and CC2420TransmitP components. This multiple wiring from one component to multiple components is called *Fanout wiring*. Correspondingly, there is the opposite situation where multiple callers are calling one callee as in the example below, which is known as *fanin wiring*.

Fanin Suppose that a command *c* is invoked from component A and component B, where c is implemented in an interface implemented in another component C. (See code below.) This is an example of fanin. When *C* issues a corresponding event signal indicating the completion of the execution of the command *c*, the event-handling functions of both *A* and *B* are invoked, and as before, the order in which they are invoked is unknown. An

```
{
  components A, B, C;
  A. IF -> C. IF;
  B. IF -> C. IF;
}
```

interface may be used by more than one user(fan in). Similarly, an interface may be provided by more than one provider(fan out).

PROBLEMS

6.1 Explain the concepts of fanin and fanout.

6.2 Summarize the notion that tasks represent in TinyOS.

6.3 What is an asynchronous command?

6.4 What issues arise when atomic blocks are improperly used?

6.5 Write a simple application to continually increment a counter value and send to another mote where the process is repeated.

6.6 Refer to the interface folder in theTinyOS system: tos/interfaces.

6.7 Examine the interface Leds.nc, and find out what each command does.

6.8 Consider the interface Boot.nc. Why does it have only one event function?

6.9 Consider the interfaces StdControl.nc and SplitControl.nc. What difference do you find between these two interfaces?

6.10 Provide a complete implementation for the data type stack, and show its wiring diagram.

6.11 Consider a simple monitoring application where we use two nodes n_1 and n_2. Each node (e.g., n_1) when sensing an object within its range raises an alarm (by turning its red light ON), and at the same time sends an alert message to the other node (n_2), which turns its green light ON. We need to implement this in nesC. Give the interface descriptions, and show the implementation of each interface function. Also, show the final overall wiring diagram.

REFERENCE

1. P. Levis and D. Gay, TinyOs Programming, Cambridge Univ. Press, 2009.

PART III
Sensor Network Implementation

7 Sensor Programming

Programming is an artform that fights back.

—Chad Z. Hower

As sensing devices become more complex, applications are being developed that exploit the increased power these devices provide. Traditional programming models and abstractions only inadequately capture these newly found evices, and thus there is a need to develop different types of techniques that naturally exploit the power of these sensing devices. Sensor programming models should essentially support abstractions over collections of sensor devices, heterogeneous data emanating from these devices, and their data storage capabilities. A sensor network can be viewed from several angles. Our view is committed to a programmer's perspective where a programmer would like to program the nodes in the network at a fairly detailed level, and at the same time is interested in programming the network at the abstract group (cluster) level. At the node level, a sensor can be viewed as a sensing device that senses the environment measuring, for example, temperature, pressure, and smoke intensity. At the higher level, we view a collection of nodes as a distributed source of data streams.

7.1 PROGRAMMING CHALLENGES IN WIRELESS SENSOR NETWORKS

One of the main challenges in wireless sensor programming involves updating and managing several resource-constrained nodes that interact in real time with the physical world, where these nodes cannot be easily accessed. We first discuss some of the system interfaces available in TinyOS [2].

7.1.1 System Interfaces

In the examples in earlier chapters, we defined our own interfaces, provided implementation modules, and wired configurations. The nesC [2] language is meant primarily for programming embedded systems such as motes (tiny devices) in wireless sensor network applications. The PrintMessage program we wrote in Chapter 6 is hardly useful in nesC applications, since such programs could be written much

Fundamentals of Sensor Network Programming: Applications and Technology, By S. S. Iyengar, N. Parameshwaran, V. V. Phoha, N. Balakrishnan, and C. D. Okoye

more comfortably in C. So, if we want to use the sensing and the wireless transmission capability of the motes, we need to get acquainted with the rich collection of built-in interfaces, library of modules, prewired configurations, and predefined types supported in nesC. In order to acquaint ourselves with some of the simple interfaces, modules, and configurations, let us begin with an example provided in the TinyOS package, called *PowerUp*. This program uses two interfaces, Boot and Leds, which are already provided in the system, and does not seem to require any other interfaces.

PowerupC.nc

```
module PowerupC
{
        uses inter face Boot;
        uses interface Leds;
}
implementation
{
                event void Boot.booted()
                {
                call Leds.led0On ();
                }
}
```

PowerupAppC

```
configuration PowerupAppC{}
implementation
{
              components MainC;
              components LedsC;
              components PowerupC;

              /*The following statements denote the fact that
              the interfaces Boot and Leds used in PowerupC are
              provided(implemented) in the components MainC and LedsC
              respectively, provided in the system. */
              PowerupC.Boot -> MainC;
              PowerupC.Leds -> LedsC;
}
```

Thus, note that there are no interface declarations in this example. We are already familiar with the Boot interface. What is more interesting is the interface Leds, which specifies commands for controlling the LEDs (light-emitting diodes) mounted on the motes by the manufacturer.

The Telosb mote supplied byCrossbowTechnology supports three LEDs of colors red, green, and blue [1]. The interface Leds (see code below) provides commands for controlling each LED. Additionally, it provides commands for obtaining and setting the mask bits that are used to control the three LEDs. Since the module PowerupC (see

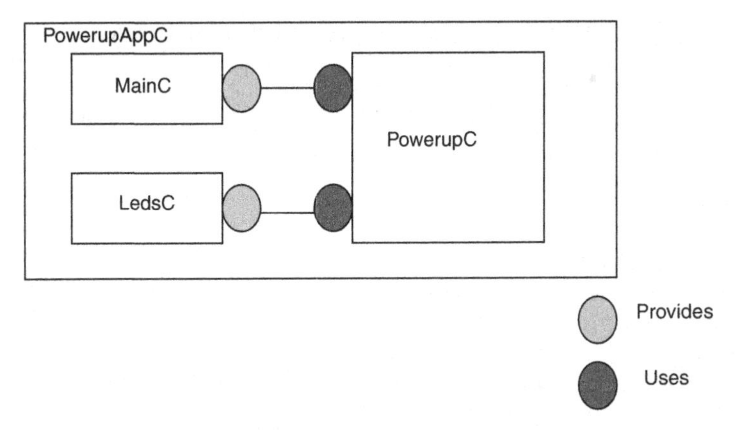

FIGURE 7.1 Component architecture for PowerupAppC.

also Fig 7.1) uses the interfaces Boot and Leds, we need to provide implementation for both of them.

```
// Courtesy of TinyOS
//authors : Philip Levis and Joe Polastre
//description : TinyOS Interface Leds
 interface Leds {

   // Turn on and off individual LEDs
   async command void led0On();
   async command void led0Off();
   async command void led0Toggle();
   async command void led1On();
   async command void led1Off();
   async command void led1Toggle();
   async command void led2On();
   async command void led2Off();
   async command void led2Toggle();

   // Get the mask bits for the LEDs.
   async command uint8_t get();

   // Set the mask bits for the LEDs.
   async command void set(uint8_t val);
}
```

In order to provide implementation for these interfaces, we browse through the modules and configurations provided in the TinyOS system, and we find that the configuration MainC provides an implementation for the interface Boot and the configuration LedsC provides an implementation for the interface Leds. We specify this implementation of the interfaces by the following wiring statements:

PowerUpC.Boot → MainC

PowerUpC.Leds → LedsC (can also be written as LedsC ← PowerupC.Leds)

The code below shows how TinyOS has implemented the interface Leds.

```
//Courtesy of TinyOS
//authors: Phil Buonadonna, David Gay, Philip Levis, and Joe Polastre
//description: TinyOS Configuration File LedsC
configuration LedsC {
provides interface Leds;
}
implementation {
components LedsP, PlatformLedsC;
  Leds = LedsP;
  LedsP.Init <- PlatformLedsC.Init;
  LedsP.Led0 -> PlatformLedsC.Led0;
  LedsP.Led1 -> PlatformLedsC.Led1;
  LedsP.Led2 -> PlatformLedsC.Led2;
  }
```

We can now draw the component architecture for PowerupAppC, which contains three subcomponents, two of which were provided by the system facilitating our task of building our component PowerupAppC.

Note that the configuration LedsC provides an implementation for the interface Leds by further wiring Leds to LedsP using the statement

$$\text{Leds} = \text{LedsP}$$

and the interfaces of LedsP to the interfaces of PlatformLedsC, which is defined elsewhere in the system. If we really want to understand how the interface Leds is actully implemented, we need to explore the system further for explanations of PlatformLedsC, which we will not do. In order to run our program, all we need is an implementation of the interface Leds, and we have found one, namely, the configuration LedsC, which provides interface Leds. We can now run the program PowerupC. As we discussed in the previous example, when a program is run, the event *booted* is invoked (which we have given in PowerupC), which calls `led0On()` to turn the red LED on. Before we close the discussion on this example, let us revisit the wiring concept again. The wiring

$$A \cdot X \rightarrow B \cdot Y$$

means that the interface X used in component A is provided by interface Y in component B. The wiring

$$B \cdot Y \leftarrow A \cdot X$$

also means the same thing. Sometimes we also use the operator = as shown below in a configuration component C.

$$X = Y$$

This means that component C provides interface X by exporting Y, where Y is the component that it uses. Now that we have seen some system interfaces and their implementations as provided by the system, we take this opportunity to look at some more interfaces and get acquainted with them as they often may be useful in the applications that we want to develop.

7.1.2 The Timer Interface

One important interface that we will often need in our implementations is Timer.

```
// Courtesy of TinyOS
// authors: Cory Sharp
// description: TinyOS Timer Interface
interface Timer<precision_tag>
{
        command void startPeriodic(uint32_t dt);
        command void startOneShot(uint32_t dt);
        command void stop();
        event void fired();
        command bool isRunning();
        command bool isOneShot();
        command void startPeriodicAt(uint32_t t0, uint32_t dt);
        command void startOneShotAt(uint32_t t0, uint32_t dt);
        command void command uint32_t getNow();
        command void command uint32_t get t0();
        command void command uint32_t getdt();
}
```

This is particularly useful in the context of wireless sensor programming, where we have to be aware of time, and this is achieved by starting a clock at the beginning and managing it from then on. Since batteries have limited lifetime, we need to save energy by avoiding unnecessary transmissions, particularly when other nodes are not listening. Thus, there is a need to go into sleep mode as often as possible, and when the sensor node wants to go into sleep mode, it has to make sure that other nodes are not trying to interact with it by, for example, trying to send some data or waiting for data to be sent. This requires planning activities well in advance so that everything is scheduled and executed in the planned way, and we need a timer for this. In fact, there are two important resources that a node must share with the other nodes: time and the medium. The Timer interface provides all basic operations that we need to manage the time within a node. The Timer is also an example of parameterized interfaces. In the Timer interface declaration *precision tag* is a type parameter that can be instantiated to give the timer the required precision. For example, TinyOS defines the following types:

```
typedef struct { int notUsed; } TMilli;
typedef struct { int notUsed; } T32khz;
typedef struct { int notUsed; } TMicro;
```

These types define the precision of a clock that we want. `TMilli` refers to a precision of one millisecond (1-kHz clock), `T32khz` refers to a 32-kHz clock, and `TMicro` refers to a microsecond clock.

Sometimes in our programs we will need a regular clock that, when execution starts, begins running as any real world clock would do. Such clocks are implemented using the startPeriodic command. Suppose that we have an interface `Timer<TMilli>`, then `startPeriodic(250)` of this timer will start a 250-kHz clock that fires 250 $\times$ 1023 pulses every second starting from now until this timer is stopped, where as `startOneShot(250)` fires only one pulse after 250 milliseconds, starting from now, and then stops.

An operation that is important to sensor programming is provided by the event `fired()`. Whenever a timer fires a pulse, the TinyOS system invokes the event `fired()` to signify the fact that time has advanced by a unit of time. We will be using this feature often whenever we use the Timer interface. This also, of course, means that we need to implement the event function `fired()` since we use the interface. Sometimes it is useful to start the timer after some delay, and we can do this using the command, for example, `startPeriodicAt(5,250)` which starts a periodic timer with a precision of 250 milli-seconds beginning 5 milli-seconds from now. Other commands are described in the system, and we will illustrate how to use some of them in the examples below. Fortunately, the Timer interface is already implemented, so it will suffice if we learn how to call its various commands and the event.

Given below is a simple application where we have used one command function and the event function from the interface Timer. As we know by now, an application will consist of interface definitions, their implementations using modules, and configurations showing how the components are wired together. In this example we use three interfaces Timer, Leds and Boot, and all these interfaces have already been implemented in the system, so our only task is to implement only the events defined in these interfaces.

```
module TwinkleC {
uses interface Timer<TMilli> as MyTimer;
uses interface Leds;
uses interface Boot;
}
implementation {
event void Boot.booted() {
call MyTimer.startPeriodic( 200 );
   }
event void MyTimer.fired() {
call Leds.led0Toggle();
 }
}
```

This program describes a module Twinkle, where we have shown an implementation for the event function `booted()` from the interface Boot, and the event function

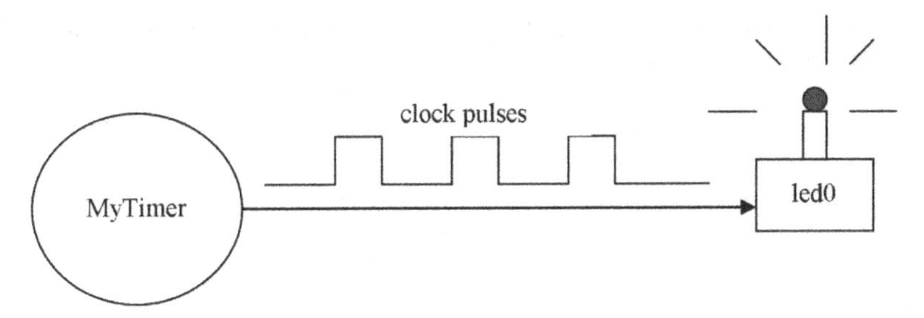

FIGURE 7.2 An illustration of where Timer drives led0.

`fired()` from the interface Timer, and have inserted calls to commands that toggle the red LED. Let us now explain how the control flow occurs in this code. To begin with, note that we have chosen a time of millisecond precision and renamed it as MyTimer. When we run this application, the TinyOS system will begin its execution by calling the event function `booted()`. Inside `booted()`, we call the command function from the interface MyTimer `startPeriodic(200)`, which starts a clock that fires a pulse every 200 milli-seconds. Thus the control flow of execution that started with `booted()` has jumped to `startPeriodic(200)`. As we have said before, whenever a clock pulse is fired, the event function `fired()` from Timer is called, and thus the control has flown now to the event function `MyTimer.fired()`, where we call the command function `led0Toggle()`, which toggles the red LED. When the clock fires its next pulse after another 200 milli-seconds, the red LED toggles again. This continues indefinitely, as we have illustrated in Fig 7.2. We now can complete the application by specifying the components that provide the interfaces that we have used in the module TwinkleC, and this is done in TwinkleAppC which shows the required wiring. As we know, LedsC provides (the implementation for) the interface Leds, and MainC provides the interface Boot. We should now also remember that TimeMilliC provides an implementation for the Timer interface. Note that this has been renamed as SystemTimer0.

```
configuration TwinkleAppC {
components MainC, TwinkleC, LedsC;
components new TimerMilliC() as SystemTimer0;
TwinkleC.Boot -> MainC.Boot;
TwinkleC.Timer0 -> SystemTimer0;
TwinkleC.Leds -> LedsC;
}
```

7.2 SENSING THE WORLD

We will now look at one application where we sense the world using the sensor mounted on the mote, and turn on the LED if the data we read have a 1 as the least significant digit. This contains a module SenseC, an interface Read, and a configuration SenseAppC.

```
// Courtesy of TinyOS
// authors: Jan Hauer
// description: Sense Application \label{prog7}
module SenseC
{
uses
{
       interface Boot;
       interface Leds;
       interface Timer<TMilli>;
       interface Read<uint16_t>;
}
}
implementation
{
       event void Boot.booted() {
       call Timer.startPeriodic(10);
}
 event void Timer.fired()
 {
       call Read.read();
 }
 event void Read.readDone(errort result, uint16_t data)
 {
       if (result == SUCCESS)
       {
               if (data & 0x0001)
                   call Leds.led0On();
               else
                    call Leds.led0Off();
       }
 }
```

SenseC This module uses four interfaces, three of which we have already seen: Boot, Leds, and Timer. (See code above.) The fourth interface, Read is described below.

Read This interface provides one command function and one event function to read the sensor and return the value read.

```
interface Read<val_t> {
command error_t read(); {
  event void readDone(error_t result, val_t val);
  }
```

The command `read()` is invoked for the read operation. We can access the results of the read command by calling the event function `readDone(result,val)`. When we call as `readDone(result,val)`, we can retrieve the value of the data read in the parameter val if the result of reading was successful, i.e., if result = SUCCESS. Otherwise, parameter `val` may contain arbitrary values.

It is now easy to understand the SenseC module, and we now specify the wiring. We need to know that Read is implemented in DemoSensorC (providedbyTinyOS), an instance of which has been renamed as Sensor.

SenseAppC This is the configuration that shows the wiring required. As we did before, the interface Boot is implemented by MainC, Leds by LedsC, Timer by TimerMilliC, and Read by Sensor.

```
// Courtesy of TinyOS
// authors : Jan Hauer
// description : Sense Application Configuration
 configuration SenseAppC {
 }
 implementation {
 components SenseC, MainC , LedsC, new TimerMilliC(),
 new DemoSensorC() as Sensor;
   SenseC.Boot -> MainC;
   SenseC.Leds -> LedsC;
   SenseC.Timer -> TimerMilliC;
   SenseC.Read -> Sensor;
   }
```

The StdControl interface (defined in TinyOS as shown in StdControl code below) is an important interface that can be used to turn on and off a component C that provides (an implementation of) this interface. The start command can be used to turn on the component C and the stop command to turn it off.

Finally, we have shown in Figure 7.3 the flow of control in the nesC program execution as the application executes. The TinyOS system invokes the even booted() which calls the command StartPeriodic(). The StartPeriodic command invokes the event fired() which calls the command read () implemented in the Read module of the System. When the data is read by the read() command, readDone() event is invoked which calls the command leld0on() to turn the light on.

```
// Courtesy of TinyOS
// authors: Kevin Klues and Joe Polastre
// description: TinyOS StdControl Interface
 interface StdControl {
 command error_t start(); {
 command error_t stop();
      }
```

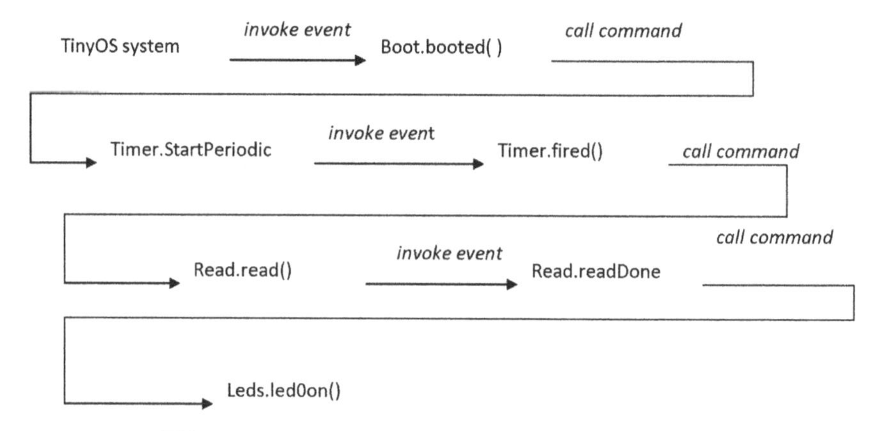

FIGURE 7.3 Control flow shown using an AND-OR graph.

7.3 APPLICATIONS USING THE INTERFACE SplitControl

SplitControl is an interesting interface that provides an opportunity for the programmer to write programs in an embedded system where a physical hardware system and its equivalent software-simulated system can be used interchangeably. Accordingly, SplitControl can also be viewed as an extended concept of StdControl and as the split-phase counterpart to the StdContol interface. The TinyOS programming manual [3] delves into more details.

```
//Courtesy of TinyOS
//authors: Kevin Klues and Joe Polastre
//description: TinyOS SplitControl Interface
interface SplitControl
{
      command error_t start();
      event void star tDone(error_t error);
      command error_t stop();
      event void stopDone(error_t error);
      event void stopDone(error_t error);
}
```

7.3.1 Sensing the Temperature

In this application, we show how we can read the temperature using a sensor node (mote) and turn an LED on.

SightC The module SightC uses five interfaces, and thus it needs to provide implementations for all the events that are declared in those interfaces. Every function defined in this module implements an event except the last function, errorReporter, which is called locally to turn the red LED on/off. We need to find an implementation for each one of these interfaces. We already know which system components provide implementations for Boot, Timer, and Leds. Reading the temperature sensor will require device-dependent command functions, so let us commit ourselves to the TelosB mote (Crossbow Technology). The TinyOS system provides a component named `Msp430InternalTemperatureC()`, which provides an implementation for the Read interface. We can directly wire to this component, but for the sake of readability let us define a configuration called TemperatureC to provide the interface Read. This is done simply by exporting `Msp430InternalTemperatureC.` Read as Read using the = wiring operator. Note that Read is declared in the statement provides in TemperatureC.

```
generic configuration TemperatureC() {
 provides interface Read<uint16_t>;
}
implementation {
 components new Msp430InternalTemperatureC();
 Read = Msp430InternalTemperatureC.Read;
}
```

We now provide the code for the module SighC.

```
module SightC
{
  uses
  {
    interface Boot;
    interface SplitControl as SightControl;
    interface Timer<TMilli>;
    interface Read<uint16_t>;
    interface Leds;
  }
  implementation
  {
  void errorReporter(int error);
  uint16_t sensor_value = 0;
  event void Boot.booted()
  {
      errorReporter(1);
      if((call SightControl.start()) != SUCCESS)

              errorReporter(1);
  }
  event void SightControl.startDone(error_t err)
```

```
  {
     if(err == SUCCESS) ({
     {
      errorReporter(0);
      call Timer.startPeriodic(TIMER FREQUENCY);
     }
     else
     {
              errorReporter(1);
     }
}
event void SightControl.stopDone(error_t err) { }
event void Timer.fired()
{
        call Leds.led2Toggle();
        if((call Read.read()) != SUCCESS)
        errorReporter(1);
}
event void Read.readDone(errort result , uint16_t val)
 {
       if(result == SUCCESS) {
               sensor value = val;
               call Leds.led1Toggle();
       }
       else
       {
        errorReporter(1);
       }
       }
       void errorReporter(int error) {
       if(error == 1) {
          call Leds.led0On();
       }
       else     {
         call Leds.led0Off();
          }
 }
}
```

Finally the following code for the configuration shows how wiring is done.

```
configuration SightAppC { }
implementation
{
```

```
    components MainC;
    components ActiveMessageC;
    components new TimerMilliC() as Timer0;
    components new TemperatureC() as Sensor;
    components SightC;
    components LedsC;
    SightC.Boot->MainC;
    SightC.SightControl->ActiveMessageC;
    SightC.Timer->Timer0;
    SightC.Read->Sensor;
    SightC.Leds->LedsC;
}
```

7.3.2 PacketSender

This is a simple example where we send a packet of data from one mote (sensor node) to another mote and also illustrate how to use the three intrfaces Packet, AMsend, and Receive. This is the first example where we attempt communication from one node to another. Fig 7.4 shows node A transmitting a packet of data to node B. In the program we discuss below one node creates a packet consisting of an integer number, sends it to a neighboring node and waits for a message from it. (We load a copy of this application program in each node.) It uses the LEDs to display the intermediate states of execution indicating the end of subtasks that are achieved successfully as the application continues to run.

It should by now have become clear to us that we will be needing the following interfaces: Boot to boot (start) application, Leds to display the intermediate states of the execution (LEDs are the only output devices we have for the human to look at and understand what is happening inside the sensor nodes), and Timer to run a clock that can generate clock pulses to fire events. We will also need interfaces for communication purposes. The interface AMSend can be used for sending and Receive for receiving packets of messages. We need one interface that can be used to manipulate with packets, and this can be done using the interface Packet. We would also like to have the turn on and turn off control on the transmission/receiving hardware components, so we will employ the interface SplitControl for this purpose. Module PacketSenderC shows nesC code that implements the packet sender. Note that SplitControl has been renamed as PSControl.

FIGURE 7.4 Node *A* sending a packet to Node *B*.

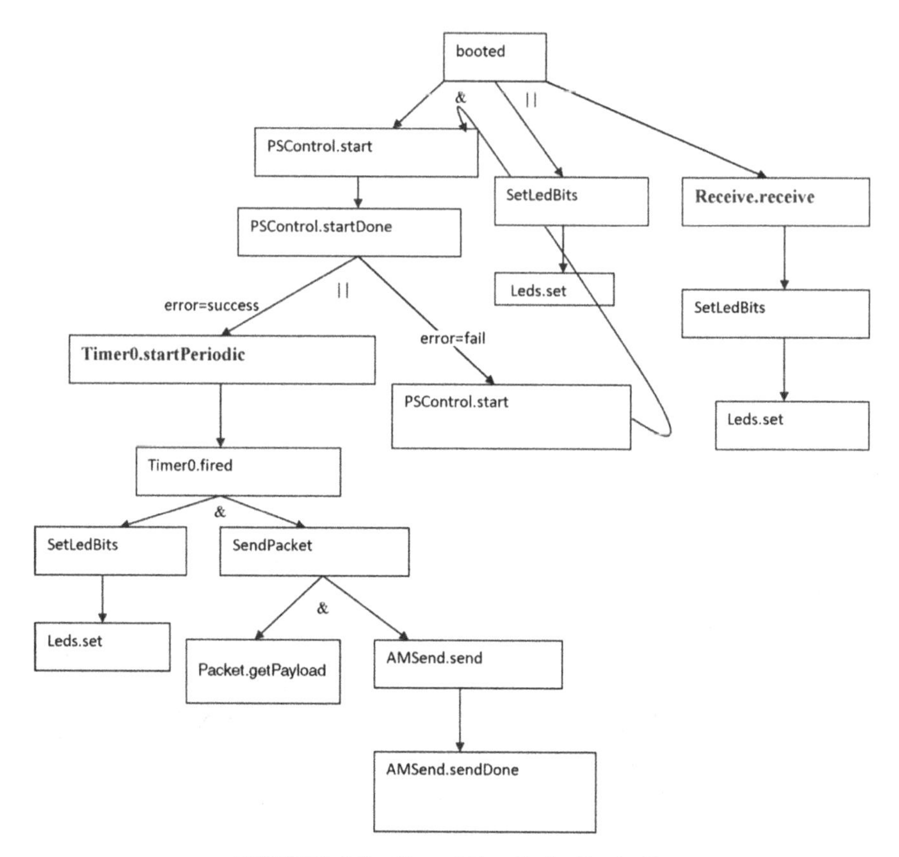

FIGURE 7.5 Control flow PacketSenderC.

The module shows the list of interfaces used, a few global variables declarations, and the prototypes of the functions that are defined in the module. In order to understand how the module works, let us trace the control flow of execution starting from the event function booted() from the interface Boot. See Fig 7.5 where we have depicted the call sequence of the command and even functions as the module starts execution.

Solid arrows show one thread of control and the dotted arrow the other. Let us start with the top most node in the graph, namely, booted. As we may recall, the TinyOS system begins its execution at the event function Booted.booted (). (As we trace down in the figure, the reader may also want to verify the control flow in the program module.)

Boot.booted() invokes PSControl.start() to turn on the transmission/receiver hardware component which attempts to turn on the component and returns the results in the parameters of the event function PSControl.startDone(). The event function startDone() checks to see if there has been any error in starting the device. If there was any error, we go back and call the start() function again. However, if the starting of the device was successful(that is, error=success), we start the clock by calling

Timer0.startPeriodic() which invokes the event function Timer0.fired(). (See implementation for event PSControl.startDone() below.) As soon as this happens in the function fired(), we turn the LEDs (by setting the bits to 0011) on showing that we were able to successfully start the transmission device. (See left branch in Fig 7.5.) Then we proceed to send a packet (calling sendPacket()) by first constructing the packet (calling Packet.getpayload()), and then sending it by calling AMSend.send(). The command send() upon completion of its task invokes the event function AMSend.sendDone which verifies if the packet was successfully sent. (See code for event void AMSend.sendDone() below.) We then go back and turn on the LEDs to indicate that the packet was successfully sent. (See middle branch under the node labeled booted.) Also, note that this part of the code appears in the event function Boot.booted().) In the meantime, if the node had received any packets, the event function Receive.receive() is invoked where we turn the LEDs on (setting the bits to ox7) to indicate that a packet has been received successfully.

```
module PacketSenderC
{
 uses interface Boot;
 uses interface Leds;
 uses interface Timer<TMilli> as Timer0;
 uses interface AMSend;
 uses interface Receive;
 uses interface SplitControl as PSControl;
 uses interface Packet;
}

implementation {
 uint16_t counter =0;
 message_t pkt;
 bool busy = FALSE;
 error_t err;
 void setLedBits(uint8_t state);
 void sendPacket();
/******************************************
Bit mask corresponding to status of LEDs.
0001 - Booted.
0011 - Trying to send Packet.
0101 - Sent Packet.
0111 - Received Packet.
******************************************/
```

We have used a set of mask bits to choose which LEDs to turn on to signify the intermediate states of the PacketSender program. Thus, after booting, we turn on the LED (mask bits are 0001), then turn on led0 and led1 (mask bits are 0011) to signify the fact that we are trying to send a packet, and so on. The timer is set to a predefined value TIMER_FREQ.

```
// module PacketSenderC (continued)

event void Boot.booted() {
       call PSControl.start();
       setLedBits(0x1); // Packet was sent successfully.
 }

event void PSControl.startDone(error_t error) {
       if(error == SUCCESS) {
             call Timer0.startPeriodic(TIMER_FREQ);
       }
       else  {
             call PSControl.start();
       }
 }

event void Timer0.fired() {
       dbg("x", "Timer Fired\n");
       setLedBits(0x5); // 0x5 = 0011
       sendPacket();
 }

void setLedBits(uint8_t state) {
       call Leds.set(state);
 }

void sendPacket() {
      BlinkData* myBD;
      counter++;
      //create a packet, then send it
      myBD = (BlinkData*)(call Packet.getPayload(&pkt, NULL));
      myBD->count = counter;
      if((err =
                     (call AMSend.send(AM_BROADCAST_ADDR,&pkt,
                      sizeof(BlinkData)))))
      {
            dbg("x", "AMSend reports success\n");
            busy = TRUE;
      }
      else if(err == FAIL)
      {
            dbg("x", "Error Could not send packet\n");
```

```
        }
}
event void PSControl.stopDone(error_t error)
 {//does nothing for now
 }
```

We have not had any opportunity to invoke PSControl.stop(), so in this implementation there was no need to invoke the event function PSControl.stopDone(). A counter, initialised to 0, has been used to provide the data in the packet being sent. The dbg function prints messages on the command window if the program is run on the TOSSIM similator. When we run this program on the sensor nodes, the dbg output is ignored. The boolean variable *busy* is to indicate whether the sender is busy or not.

```
   // module PacketSenderC (continued)

 event void AMSend.sendDone(message_t* msg, error_t error) {
        if((&pkt == msg) && (error== SUCCESS))
        {
               dbg("x","Packet was sent successfully\n");
               busy = TRUE;
        }
        else
        {
               dbg("x", "Packet was not sent\n");
        }
 }

 event message_t* Receive.receive(message_t* msg,
                                    void* payload, uint8_t len)  {
        setLedBits(0x7); // received data successfully.
        dbg("x", "Received a packet\n");
        return msg;
 }
}
```

PROBLEMS

7.1 What is the role of timers in wireless sensor applications?

7.2 In every TinyOS-based application, what interface is responsible for system initialization?

7.3 What are some of the challenges of writing wireless sensor applications?

7.4 What role does the SplitControl interface play in TinyOS?

7.5 List the differences between SplitControl and StdControl interfaces.

REFERENCES

1. *Crossbow Telosb Datasheet,* Courtesy Crossbow Inc.
2. J. Hill, R. Szewczyk, A. Woo, S. Hollar, D. Culler, and K. Pister, System architecture directions for networked sensors, in *In Architectural Support for Programming Languages and Operating Systems,* 2000, pp. 93–104.
3. P. Levis and D. Gay, *TinyOs Programming,* Cambridge Univ. Press, 2009.

8 Algorithms for Wireless Sensor Networks

Talk is cheap. Show me the code.

—Linus Torvalds

An *algorithm* can be defined as a logical sequence of instructions for solving a problem in a finite sequence of steps. In wireless sensor networks, the design of algorithms becomes an important issue as energy and computational resources are scarce and therefore must be effectively put to good use. With the limited computational ability of each individual node, multiple sensors nodes collaborate to solve tasks using complex parallel processing techniques. These parallel processing techniques rely on efficient parallel algorithms to achieve collaboration. Therefore, in this context, a *parallel algorithm* can be defined as an algorithm in which several computations are carried on simultaneously across multiple processing units. In order to extend the life of a sensor network, these parallel algorithms have to be developed to be efficient in network resource usage; hence, the need for energy-aware networking algorithms [7,10,9]. There are numerous networked computing devices of all shapes and sizes from handheld computers such as personal digital assistants and mobile phones, to more powerful systems such as laptops, desktop workstations, and supercomputers. Most of these devices communicate over traditional IP-based networks supporting huge data transfer rates due to rapidly increasing band width and faster processing capabilities. These networks, being highly scalable and structured, may consist of several routers, switches, and bridges interconnecting millions of nodes and may use complex routing schemes to transfer data from one end device to another. With energy and computational resource usage being of no consequence (energy can be replenished for most devices), communication between multiple devices is very cheap and reliable. It is in this respect that wireless sensor networks chiefly differ from conventional networks. With nonrenewable sources of power and very little onboard power, computation and communication come at a very high price. As wireless sensors become more pervasive because of their lower cost, we can expect the proliferation of sensors to exceed those of traditional computing devices. Hence, there is a need to develop new energy-aware routing algorithms and aggressive power

Fundamentals of Sensor Network Programming: Applications and Technology, By S. S. Iyengar, N. Parameshwaran, V. V. Phoha, N. Balakrishnan, and C. D. Okoye

management schemes for this newly emergent class of computing devices [4]. In this chapter, we will discuss several concepts, challenges, properties, and algorithms unique to wireless sensor networks, such as

- Communication patterns prevalent in sensor networks
- Physical components of sensor nodes
- Properties of wireless sensor networks
- Networking layers
- Routing

Also, we will show how some of these concepts can be implemented using nesC code and pseudocode.

8.1 STRUCTURAL CHARACTERISTICS OF SENSOR NODES

The term *sensor nodes* (also called *motes*) refers to a sensing device that belongs to a wireless sensor network that is capable of processing, gathering, and communicating sensory information with other devices in the network. The architecture of a typical mote is shown in Fig. 8.1.

With advances in microelectromechanical systems (MEMSs) and low-power wireless technology, the major components of sensors as shown in Fig. 8.1 have been miniaturized over the years. However, the processing and storage capabilities of sensors have not developed in accordance with Moore's law. In the following section, we compare the properties of some of these components (microcontroller, transceiver, memory, power unit, and sensors) with their traditional counterparts.

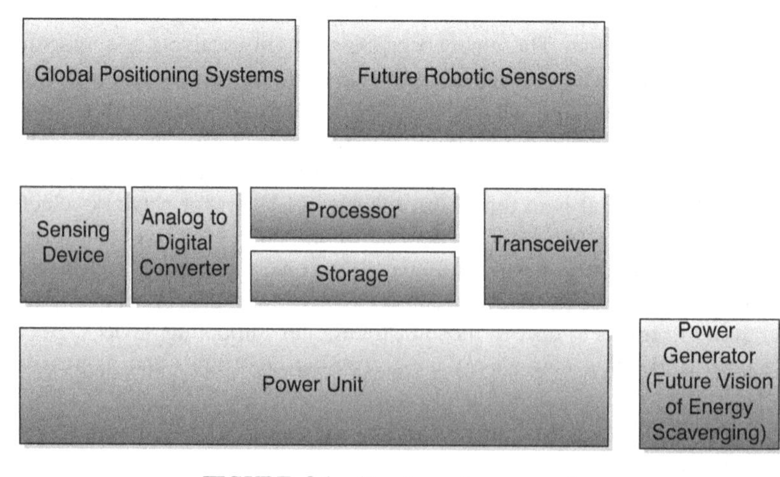

FIGURE 8.1 Structure of sensor nodes.

8.1.1 Microcontroller

The microcontroller is one of the most important components in a sensor.It is responsible for data processing, memory management, interrupt handling, and the control of other components on the sensor to minimize energy usage and increase the lifespan of the motes. Unlike the microcontrollers present in traditional computers, most sensors operate on less than a watt of power, giving them a lifespan of several months to a year. This power conservation is achievable because of the ability of microcontrollers to enter low-power states when idling. These microcontrollers have frequencies between a few kilohertz and several megahertz, allowing for the most basic tasks such as sensing and intermediate routing of data from other sensors.

8.1.2 Transceiver

Currently, information can be transmitted in wireless sensor networks through three types of media:

- Optical communication (laser)
- Infrared (IR) communication
- Radio frequency (RF)

Although both optical and IR communications have the advantage of requiring less energy to transmit data, they both suffer from major drawbacks. For instance, optics-based communications require both communicating nodes to be properly aligned in direct line of sight. Conversely, IR communications have very short ranges. Other issues include sensitivity to atmospheric conditions in optical communications and unidirectional communication for both systems. It is for these reasons that RF is more prevalent in sensor networks.

8.1.3 Memory

As noted in earlier chapters, memory is a scarce resource on sensor nodes. Therefore, applications must be efficient in terms of not only energy but also the amount of memory they use. For instance, a mica2 mote contains 644 kb of memory that must be used by the TinyOS operating system and applications built on the platform [1].

8.1.4 Power Unit

Most power consumption in sensor nodes is primarily for data processing, sensing, or communication operations. Of these operations, communication activities exert the most toll on the battery life of the sensor. It is for this reason that new low-power components are being created and power-saving policies, such as dynamic power management (DPM) and dynamic voltage scaling (DVS), are continually being refined to provide ever-increasing energy savings. A typical sensor node is shown in Fig. 8.2.

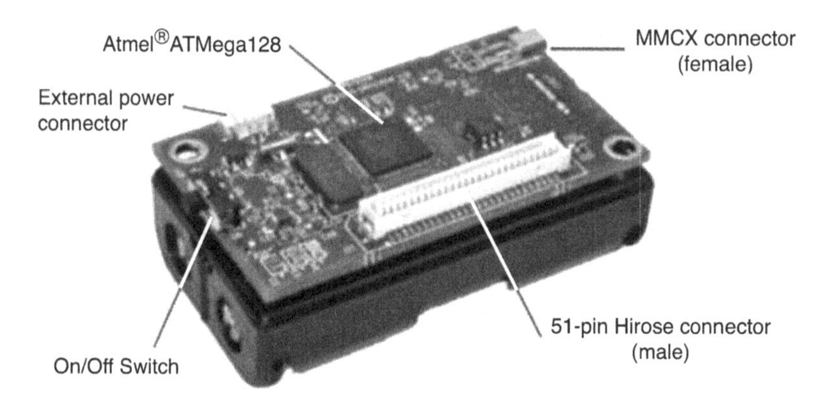

FIGURE 8.2 A typical sensor node (mote).

8.2 DISTINCTIVE PROPERTIES OF WIRELESS SENSOR NETWORKS

With the unpredictable nature and unsuitability of traditional routing algorithms in wireless sensor networks, newer, more efficient, and more reliable algorithms have been developed. These algorithms were developed to address the challenges posed by the volatile wireless communication in sensor networks. These challenges stem from the ubiquitous and heterogeneous nature of WSNs, making central management of individual sensor nodes nearly impossible. As a result, intelligent features such as self-configuration, healing, dynamic routing, and multihop communication abilities have been suggested to improve the reliability and management tasks in these networks. In the following section, we discuss each of these attributes and how they enhance communication in wireless sensor networks.

8.2.1 Self-Configuration

Wireless sensor networks are usually composed of thousands of nodes randomly deployed and organized in order to achieve similar objectives (see Fig. 8.3 for an example). These objectives are to retrieve certain data about an entity being monitored and transmit results to a remote destination serving as a data sink. The random deployment of sensors and the volatile nature of the network (usually due to node failure) warrant the development of self-configuration mechanisms to prevent network degradation and efficient transmission of information.

8.2.2 Self-Healing

Despite measures taken to ensure durable sensor networks, several factors still exist that can result in a breakdown in communication. These factors include energy depletion in some key routing nodes, (un)intentional damage by humans or other animals, or even the addition of new nodes resulting in a highly dynamic network. It

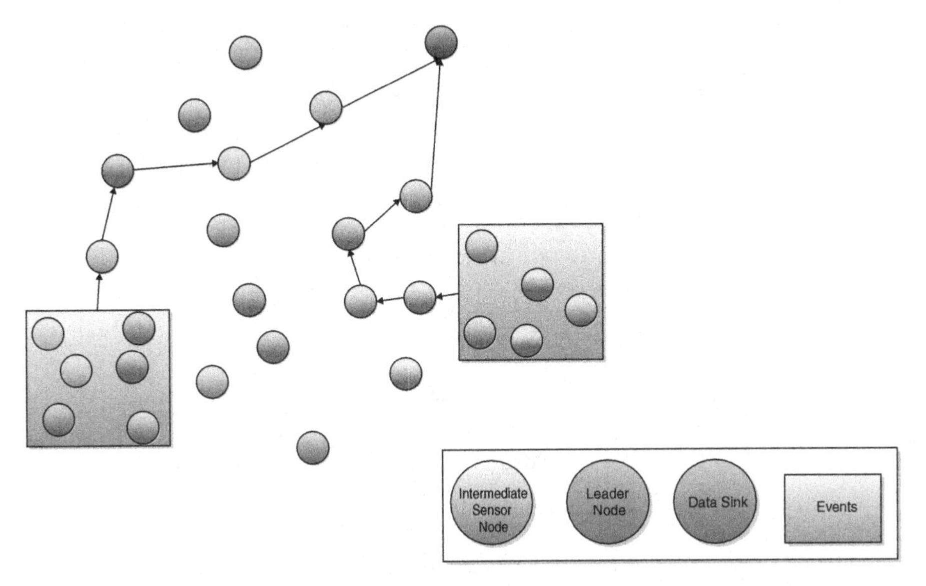

FIGURE 8.3 Dataflow in a sensor network.

is for this reason that the networking stack on sensors be impervious to node additions and deletions without requiring a complete reset of the whole network.

8.2.3 Dynamic Routing

Communication in sensor networks can be very expensive, and sensor nodes can be added and removed on the fly. The need for an adaptive routing scheme based on network conditions such as link quality, hop count, and gradients has led to the development of several new on-demand routing algorithms. These make use of lower band width for control packets, resulting in less need to food the network for periodic updates. This, in turn, results in energy savings for sensor nodes, extending the life of a wireless sensor network considerably. Examples of dynamic routing in wireless sensor networks include caching and multipath (CHAMP) routing and hierarchical state routing (HSR).

We have now examined the structural characteristics of sensors and some of the unique properties of wireless sensor networks. In the following section we will discuss the components forming the network stackin wireless sensor networks.

8.3 SENSOR NETWORK STACK

Communication, routing, and data transfer in sensor networks is possible between differing node types because of well-established standards and specifications providing low-level implementation details on how data can be exchanged in sensor networks.

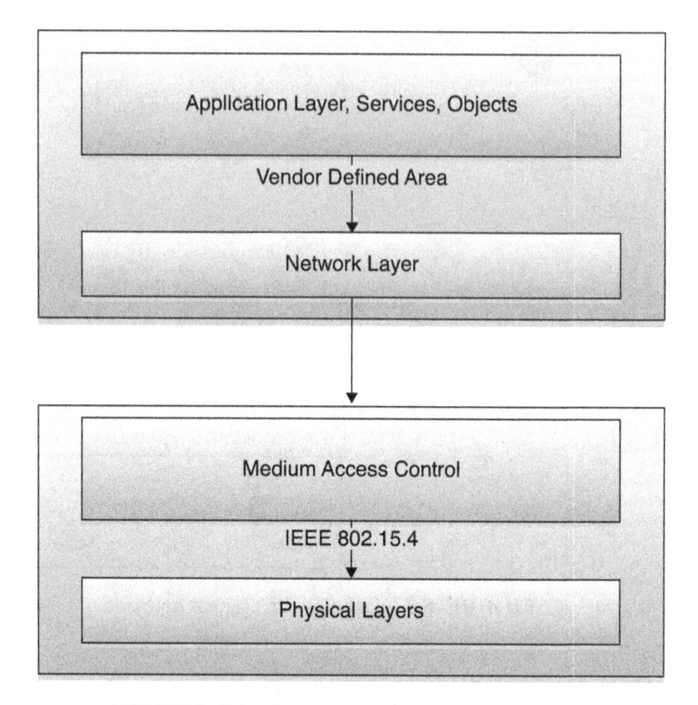

FIGURE 8.4 Sensor network stack architecture.

The IEEE 802.15.4 is the standard specifying details on data exchange in the physical and medium access control (MAC) layers for low-rate wireless networks. When sensor network–based applications are written, most interactions with the IEEE-specified layers are usually through abstract libraries implemented by independent vendors and offered through specific sensor operating systems (e.g., TinyOS). These libraries automatically transform application data into a form in conformance with IEEE specifications. These vendor-provided libraries providing software abstraction for the MAC and physical layers are referred to as the *network layer*. In this section, the several components that form the networking stack in wireless sensor networks are examined (see also Fig. 8.4).

8.3.1 Physical Layer

The physical layer is the first and lowest layer consisting of basic hardware transmission technologies of a network. It is responsible for several functions, including

- Provision of a data transmission service.
- Management of RF transceiver.
- Channel selection.

Energy and signal management functions. For wireless sensor networks, the physical layer transmits in one of three unlicensed frequency bands. In North America the most common is the 915-MHz ISM (instrument–scientific–medical) band. Among the main functions of the physical layer is the detection and correction of transmission errors. These errors could stem from several factors, such as

Attenuation—a decrease in intensity of electromagnetic energy at receiver due to long distance.

Doppler shift—a change in frequency of a wave caused by the relative velocities of the transmitter and receiver (common for mobile agents).

Hidden-terminal problem—a scenario in which the medium around the source node is free but busy around the destination node.

Exposed-terminal problem—in this case, the medium around the destination node is free but engaged at the source node.

8.3.2 Medium Access Control (MAC) Layer

The MAC layer is the second lowest layer, offering a management interface for the physical channel. It is responsible for frame validation, timeslot allocation, synchronization, and node associations in a network. There are several other implementations of the MAC Layer for sensor networks, such as S-MAC, B-MAC, and C-MAC, all of which have different strengths ranging from energy conservation to routing speed gains. References 8, 6, and 11 provide more information on MAC implementations in wireless sensor networks.

8.3.3 Network Layer

Although there is no defined standard for the network layer in sensor networks, several differing implementations exist today, the most common of which is the Zigbee specification. Details on the Zigbee network and application layer are covered in Chapter 10.

8.3.4 Full-Function Device (FFD)

Full-function devices are nodes having a general model of communication allowing them to "talk" with any other device. Also, such a device can be assigned the role of coordinator of a personal area network.

8.3.5 Reduced-Function Device (RFD)

Reduced-function devices are more restricted in their functions. They are usually very simple devices with low resource and communication requirements. It is for this reason that they can communicate only with FFDs and can never become network coordinators.

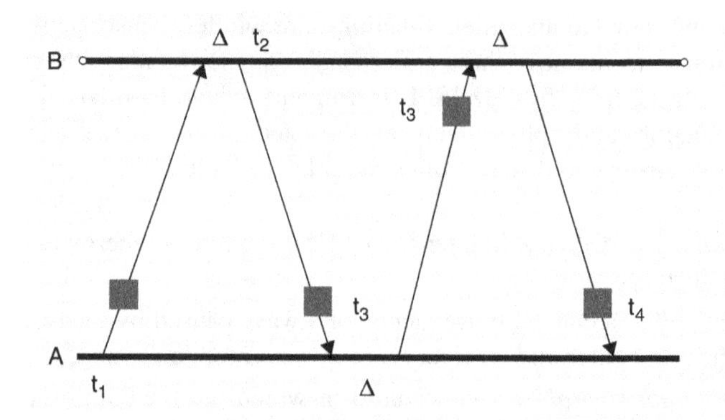

Equations

$t_2 = t_1 + \text{delta} + (\text{propagation\_delay})$
$t_4 = t_3 + \text{delta} + (\text{propagation\_delay})$

FIGURE 8.5 Receiver–receiver synchronization.

8.4 SYNCHRONIZATION IN WIRELESS SENSOR NETWORKS

Time synchronization refers to a method of timekeeping requiring the coordination of multiple events to operate a system in unison. It is particularly important in computing because of its critical role in communication. Distributed systems rely on several synchronization mechanisms to ensure proper operation without which severe degradation in performance can occur [5, 2, 3]. In data aggregation and event monitoring sensor networks, the need for a flexible and robust time synchronization mechanism is more apparent since collaboration among nodes is critical to data reduction and the energy efficiency of sensor networks. Also, sensor usually monitor highly dynamic environments that change with time. For this reason, time is a basic requirement for nodes to correlate events that occur in the network with events in the physical world. Failure to synchronize sensor nodes in a sensor network would render measurements useless since synchronization is essential for both temporal and spatial analysis of events (most tasks fall into these two categories, such as node localization, and target tracking). Several methods exist to synchronize nodes in a network, the most basic of which is sender–receiver or receiver–receiver synchronization. In this approach, peers in a network conduct time synchronization using timestamps in data packets as a reference. A visual representation of the algorithm and the accompanying equations are shown in Fig. 8.5.

Some drawbacks exist with this basic method of synchronization. Most notable are the four different types of delays involved:

- *Send time*—time spent in building the packet for transmission and delays inherent in the protocol stack
- *Access time*—time spent waiting for the medium to become free

- *Propagation time*—time taken for the packet to reach the destination
- *Receive time*—time spent by the receiver processing the packet before it is timestamped

Although send and access time delays can be eliminated with the use of reference broadcast synchronization, some newer approaches exist that are less error-prone.

We will now discuss the programming challenges involved in implementing a few basic algorithms [4] from the WSN area. First, the beaconing behavior of surrounding nodes is examined.

8.4.1 Beaconing

Beaconing refers to the continuous transmission of small control packets that notify neighboring nodes about the presence of the transmitter. The pseudocode describing how beaconing works is presented below.

```
begin:
        while (true):
                broadcast (address, random time);
                sleep (random time);
end
```

As we did before, we assume that the node has the three basic capabilities (sense, broadcast and sleep) as enumerated above. In this example, we use a procedural style of code

Procedural Beaconing

```
begin:
      senseMedium(signal):
           if(signal equals absent):
                  return 'no_signal'
           elif(signal equals weak):
                  return 'medium_busy_or_no_data'
           elif(signal equals collision):
                  return 'collision'
           elif(signal equals strong):
                   return 'strong'
end
```

Sometimes certain routing schemes may employ the use of lists to keep track of their nearest neighbors. A sample pseudocode and associated nesC configuration showing how a neighborhood table construction can be done is shown below.

```
{
   initialize-table(table); i =0; t1 = time-now( );
      while (i < n){
            data-packet = sense-data( );
            address = address-of (data-packet);
            if not-present(address, table)
                     {add(address, table ); i ++}
                        if  ( time-now( ) > t1 +
max-time) break;
                     }
                  }
               }
         }
```

```
#define MOTE_AM_ID 10
configuration algorithm2AppC { }
implementation
{
   components MainC;
   components RandomC;
   components LedsC;
   components new AMSenderC(MOTE_AM_ID);
   components new AMReceiverC(MOTE_AM_ID);
   ActiveMessageAddressC
   components ActiveMessageAddressC;
   components algorithm2C;
   components ActiveMessageC;
   components new TimerMilliC( );
   algorithm2C.Boot->MainC. Boot;
   algorithm2C.Random->RandomC;
   algorithm2C.Leds->LedsC;
   algorithm2C.AMPacket->AMSenderC;
   algorithm2C.AMSend->AMSenderC;
   algorithm2C.AMA->ActiveMessageAddressC;
   algorithm2C.PacketAcknowledgements->AMSenderC;
   algorithm2C.Receive->AMReceiverC;
   algorithm2C.RadioControl->ActiveMessageC;
}
```

8.4.2 Neighborhood Table Construction

In this application we use the standard interfaces shown in the program above.

Wiring for Neighborhood Table Construction The interfaces are implemented using the system-provided components shown above. The implementation details are discussed below.

```
implementation
{
      void sendSensorInformation (uint 16_t);
      void populateSensorTable (uint 16_t, uint 32_t);
      int NO_OF_MOTES = 30
      sensorInfo* sensor;
      am_addr_t sensoraddress = 0;
      uint32_t sensormac = 0x001EDD32;
      //Pre-determined mac address of current sensor
      am_group_t sensorgroup = 1;
      sensorData table [NO_OF_MOTES];
      int index = 0;
      event void Boot.booted( )
      {
          call RadioControl.start( );
      }
}

{
event void RadioControl.startDone(error_t err)
      {
          call AMA.setAddress(sensorgroup, sensoraddress);
          call Timer0.startOneShot(6000);

      }
      event void Timer0.fired ( )
   {
       message_t msg;
       sensor = (sensor Info*) call
       AMSend. getPayload (&msg );
       sensor->id = sensoraddress ;
       call AMSend. send (AM_BROADCAST ADDR, &msg,
sizeof (sensorData));
       call Leds.led0Toggle( );
      }
}
```

We will assume that there are no more than 30 motes in the neighborhood at any given time. The mote address of the given node (where we want to build the neighborhood table) is 0, its MAC address is `0x001EDD32`, and the identity of the group (which consists of the current node and all its neighbors) is 1. We begin the table construction by starting the radio.

Starting the Radio Component We set the group address to 1 and the current node address to 0, and then start a monoshot clock that invokes the event function `Timer0.fired()`, where we broadcast our node address to all neighbors and wait for their replies.

8.4.3 Implementation

Implementation Details of the Receive Method As soon as we receive a message from any neighbor, say *P*, we check whether it is a reply to the message that we sent, and if it is, we store *P*'s ID in our table. However, if the message is a request asking us to send our ID, we then send our ID to node *P*. To achieve these operations, we use the following functions: `sendSensorInformation` and `populateSensorTable`.

```
event message t* Receive.receive (message t* msg, void*
                                  payload, uint8 t len)
{
    sensorInfo* myInfo; //Specified in header file
    myInfo = (sensorInfo*)payload;
    if (myInfo->requestCode == 0x1)
    {
          call Leds.led2Toggle( ); //Toggle Led 3 to
                                    indicate info received
          populateSensorTable (myInfo->id , myInfo-
>macAddress);
    }
    else if (myInfo->requestCode == 0x0) //Send my info to some
                                          other mote
    {
         sendSensorInformation (myInfo->id);
         call Leds.led1Toggle( ); //Toggle Led 2 to indicate
                                   request received
    }
    return msg;
}

void sendSensorInformation (uint16_t receiver)
{
     message t msg;
     sensor = (sensorInfo*) call AMSend. getPayload (&msg);
     sensor->id = sensoraddress;
     sensor->macAddress = sensormac;
     call AMSend. send (receiver, &msg, sizeof(sensorData));
}
```

```
void populateSensorTable (uint16 t id, uint32 t mac)
{
     int i = 0;
     //Simple check if current mac already exists in table
     while (i <= index && table [i].macAddress != mac)
     {
           i ++;
     }
     if (table[i].macAddress == mac )
     {
            table [i].moteID = id;
     }
     else
     {
         table [++i].moteID = i d;
         table [i].macAddress = mac;
     }
}
```

Implementation Details of `sendSensorInformation` ***and*** `populateSensorTable` Additionally, we need to provide implementations for the events in our program, which are invoked when we call `AMA.setAddress`, `AMSend.send`, and `RadioControl.stop`. In `sendDone`, when we fail to send a message, we turn on an LED and then do nothing.

`stopDone`, `sendDone`***, and Changed Implementations*** In wireless sensor networks, concurrent transmission by multiple nodes could result in transmission errors due to packet corruption. An example of such a scenario would be the widely studied hidden-terminal problem. As a result of this, steps have to be taken to reduce or eliminate the occurrence of errors. The collision avoidance algorithm presented below suggests a way to avoid such errors.

```
event void AMSend. sendDone (message t msg, error t err)
{
    if (err == FAIL)
    {
            call Leds.led0Toggle ();
    }
}
async event void AMA. changed ()
{}
event void RadioControl.stopDone (error t err)
{}
```

8.5 COLLISION AVOIDANCE: TOKEN-BASED APPROACH

8.5.1 Token-Based Approach

The approach can be described very briefly using the following four abstract steps.

1. Create a token and pass it to a node.
2. If a node has the token, it transmits the data that it wants to transmit.
3. The node then transmits the token to its neighbor.
4. Use this idea to count the number of nodes in the network.

The state diagram in Fig. 8.6 captures the abstract behavior of each node in the token-based approach for collision avoidance. We implement this algorithm by building a component for it, starting with its interface.

```
module CollisionAvoidance_tokenBased
{
     uses interface Boot;
     uses interface LowPowerListening;
              interface Timer<TMilli> as Timer0;
              uses uses interface Send;
              uses interface AMSend;
              uses interface Receive;
              uses interface PacketAcknowledgement;
}
```

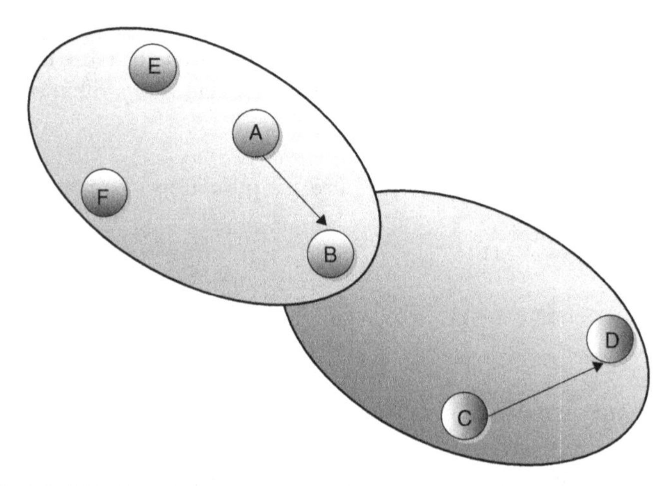

FIGURE 8.6 State diagram for token-based approach for collision avoidance.

Component Interface Definition The following program shows the wiring called *CollisionAvoidance tokenBased*. We need to build our component *boot* for booting the component, `LowPowerListening` for minimizing power consumption, `Timer0` to run a clock with millisecond precision, `send` to send packets (of messages), `AMSend` to send packets to intended destinations, `Receive` to receive packets from other nodes, and finally, `PacketAcknowledgement` to indicate whether the packet sent is to be acknowledged. For these interfaces, we choose the following implementations (as they are already provided in the TinyOS system).

```
configuration transmitAppC( )
{}
implementation
{
     components MainC;
     components Collision as AppC;
     components CC2420ActiveMessageC;
     components AMSenderC( );
     components new AMReceiverC( );
     components Timer<TMilli> as Timer0;
     AppC. Boot->MainC;
     AppC. Packet->AMSenderC;
     AppC. Send->AMSenderC; //Wrong component??
     AppC. PackageAcknowledgement->AMSenderC;
     AppC. Receive->AMReceiverC;
     AppC. Timer0->Timer0;
     AppC. LowPowerListening->CC2420ActiveMessageC;
}

Pseudocode Implementation
{
    event void Boot.booted( )
    {
          //Start one shot timer of Timer0 using
          Timer0.startOneShot (t1)
          //where t1 is sometime point in future when the
            clock fires once.
    }
    event void Timer0.fired( )
    {
          configure the LowPower Listening cycle of mote
    }
    event void Receive.received( )
    {
         configure the LowPower Listening cycle of mote
         send acknowledgement to sender;
```

```
            extract token from packet;
            transmit any data you need to
            send any node;
            relinquish the token (by sending to another node)
        }
}
```

transmitAppC For low-power listening, we wire our components to the modules that implements the CC2420 radiochip functionalities. For simplicity, we provide the implementation of our component in pseudocode.

Transmit Application Configuration File When the monoshot timer triggers have `fired()`, we put our component into low-power listening mode, and then wait for someone to send us a token. When we receive a token, we send back an acknowledgment, perform any activities that we want to perform (sending messages to any node, etc.), and once finished, we hand over the token to a neighboring node.

8.5.2 Schedule-Based Communication

In order to minimize collision and save energy, we need to follow a systematic way of transmitting and sensing behavior at each node. In this section, we write programs that follow a schedule so that the transmitter transmits only when the listener is listening, and the listener is listening only when the transmitter is transmitting. There are two ways a schedule can be embedded at each node: explicit and implicit. In the *explicit embedding*, the schedule can be described using a data structure such as a list implemented using an array. In the *implicit embedding*, the schedule is implied in the sequence of operations that the node executes. In this example, we follow the implicit style.

We can build in the schedule implicitly at each node in its program as follows. Let A:(0.B); mean at time t=0, A can transmit a packet to B. Thus, [[A:(0,B);B :(10,C); C:(20,D);D :(30,nil);E :(40,,A);F :(50,E)] can be a schedule. Total period T = 60 time units assuming that F:(50,E)] takes 10 units of time. The schedule will repeat periodically with a period T of 60 time units.

We further illustrate in this example how we can specify the behavior of the node abstractly using rules. The value of the timer, as specified by the time globally, schedules the operations so that at each node's operations are executed as per the intended schedule.

```
{
        A: t =0          transmit data packet to B.
        B: t =10         transmit data packet to C.
        C: t =20         transmit data packet to D.
        D:                NULL.
        E: t =40         transmit data packet to A.
        F: t =50         transmit data packet to E.
}
```

8.5.3 Pseudocode at Each Node

Let t stand for the timer value. (We can implement this by building a counter that goes through states triggered by a clock.)

```
module transmitC
{
    uses interface Timer<TMilli> as Timer0;
    uses interface Send;
    uses interface Receive;
    uses interface AMSend;
    uses interface Receive;
}
configuration transmitAppC
{}
implementation
{
     components MainC;
     components transmitC as AppC;
     components ActiveMessageC;
     components AMSenderC( );
     components Timer<TMilli> as Timer0;
     AppC. Boot -> MainC;
     AppC. Packet -> AMSenderC;
     AppC.AMSend -> AMSenderC;
     AppC. Send -> AMSenderC;
     AppC. Receive -> AMReceiverC;
     AppC. Timer0 -> Timer0;
}

Pseudocode Implementation
{
    event void Boot.booted( )
    {
          mode = initialize( );
          c = 0;
          Timer0.startPeriodic( );
    }
    event void Timer0.fired( )
    {
         if (mode = = transmit & counter = = 0)
         {
               AMsend( to node B, data);
               c = c+10;
         }
```

```
            if (mode == transmit & counter = = 10).
            {
                  AMsend(to node C, data);
                  c = c +10;
            }
      }
      event void Receive( )
      {
             if (mode==receive)
                  data = packet;
      }
}
```

States of Transmit The concept described above can be used to avoid collisions, as we discuss below:

- Divide time into slots.
- Insert permitted transitions at each slot. For example, the time axis is divided into intervals of period T. Each period T is divided into n slots. Thus, T = [slot 1, slot 2, ... , slot n].
- Indicate the possible transmissions in each slot. For example, [slot 1: $A \rightarrow B$, $C \rightarrow D$; slot 2:$E \rightarrow F$; slot 3:$G \rightarrow H$, $I \rightarrow J$; etc.].
- Assign to each node.

The list of interfaces we used in developing this application component are shown in the program above, along with the components that we selected for their implementation.

Schedule-Based Transmission In order to implement the schedule, we use a small counter that tells us when to transmit a packet. This counter is incremented every time a periodic clock fires. (Assume that clocks are synchronized for this implementation.)

8.6 CARRIER SENSING VERSUS DECODING

In Fig. 8.7, the green circle shows the decode range and the yellow circle shows the carrier sense. (Note that these circles are part of the manufacturer's specs.) To decode, the signal must be strong; that is, the signal-to-noise ratio (SNR) must be high. To carrier sense, the signal can be weak; that is, SNR can below.

We will now write programs for nodes B, C, D, E, F and measure the relative signal strengths.Verify with the manufacturer's specs—that is yellow circle and green circle.

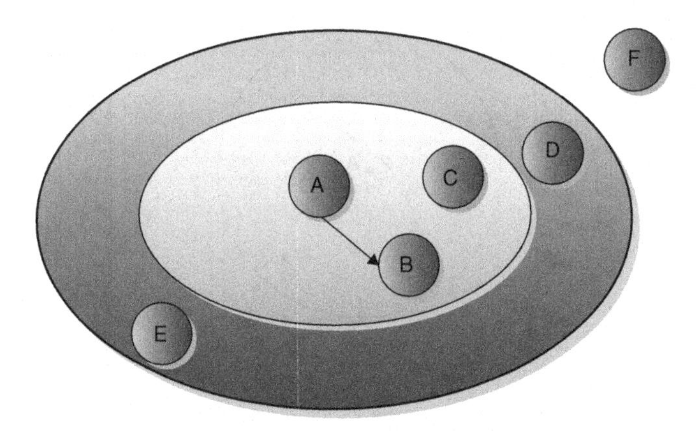

FIGURE 8.7 Illustration of carrier sensing versus of decoding.

```
begin:
         if (senseCarrier( ) equals 'NIL'):
         print ''Medium␣is␣carrier-free''
elif (senseCarrier( ) equals ''WEAK'' )
         print ''Medium␣is␣carrier-sense''
elif (senseCarrier( ) equals ''STRONG'')
         print ''Medium␣is␣carrier-decode''
end If
```

8.6.1 RTS/CTS Handshake

Ready-to-send (RTS) and clear-to-send (CTS) messages are broadcast control bytes that are exchanged between the transmitter and the receiver to coordinate the transmission of data between them. A typical handshake protocol that uses this is shown in Fig. 8.8. In this example, we also show how to convert a timing diagram into programs. Node *B* wants to transmit a data packet to node *C*. (Also worthy of note is the fact that RTS and CTS will have the MAC addresses of the sender and the receiver. Further, RTS, CTS, and the data packets will all have the duration *T* that it will take for the complete transaction. Similarly, a frame will indicate whether the transmission is unicast or broadcast.) The handshake protocol proceeds as follows:

1. Node *B* senses the medium, and goes to the next step when the medium is free.
2. Node *B* broadcasts an RTS. (Note that it is a broadcast, and not a unicast. Also, RTS contains the MAC addresses of *B* and *C*.)
3. Nodes *C* and *A* hear it. The RTS contains the total duration d_1 up to *B* receiving an acknowledgment.

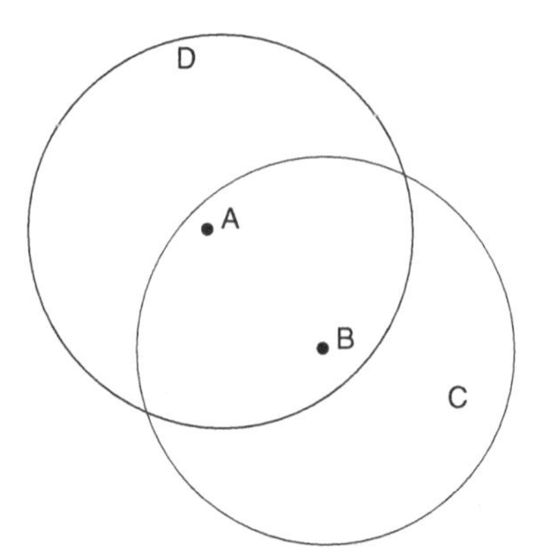

FIGURE 8.8 RTS/CTS handshake configuration.

4. Node C realizes that the RTS is for itself, senses the channel, acquires it, and broadcasts a CTS. This also contains the duration d_2. Concurrently, when A hears the RTS, it realizes that it is not for itself, but it will set the NAV flag ON (to indicate that the medium is busy) for the duration d_1.
5. When the CTS from C reaches B, B unicasts one packet of data to C. (*Note:* This unicast frame will contain the sender's and receiver's addresses, and also a flat stating that the frame is a unicast frame.) Concurrently, when D hears the CTS broadcast from C, it will set its NAV flag ON for the duration d_2.
6. Node C receives the data from B, and then sends a unicast acknowledgment to B.

We now give the pseudocode for the protocol above.

```
Node B:
begin:
        //Broadcast RTS to C: First Acquire Medium
        while (true): //Infinite loop
                if (senseMedium( ) not 'BUSY'):
                        break
                endIf
        endLoop
        //Listen for CTS from C
        while (true):
                packet <- listen( )
                if (packet equals 'CTS'):
                        break
                endIf
        endLoop
```

```
        //Send Data to C
        while (true):
                if (senseMedium( ) not 'BUSY'):
                        break;
                endIf
        endLoop
        destination <-| C
        unicast(data, destination)

        //Wait for Acknowledgement from C
        while (true):
                packet <-| listen( )
                if (packet equals 'CTS')
                        break
                endIf
        endLoop
end
Node C:
begin:
        //Listen for RTS from B
        while (true):
                packet <-| listen( )
                if (packet equals 'RTS'):
                        break
                endIf
        endLoop

        //Broadcast CTS to B
        while (true):
                if (senseMedium( ) not 'BUSY'):
                        break;
                endIf
        endLoop
        //CTS Broadcast to B
        while (true):
                packet <-' listen( )
                if (packet equals 'DATA'):
                        break
                endIf
        endLoop
        print ''Data_received_from B''


        //Broadcast ACK to B
```

```
        while (true):
                if (senseMedium( ) not 'BUSY'):
                        break
                endIf
        endLoop
        acknowledgeBroadcast(B)
end
Node A:
begin:
        while (true):
                packet <-| listen( )
                if (packet equals 'RTS'):
                        break
                endIf
        endLoop


        //B is going to send data to C, Must send A to sleep
by setting
        //NVA Variable
        duration <-| extractDuration (packet )
        NVA <-| currentTime + duration
        sleepUntil (NVA)
        print ''Woken␣up␣after␣sleeping!''
end
Node D:
begin:
        while (true):
                packet <-| listen( )
                if (packet equals 'RTS'):
                        break
                endIf
        endLoop
        //B is going to send data to C, Must send D to sleep
by setting
        //NVA Variable
        duration <-| extract Duration (packet )
        NVA = currentTime + duration
        sleepUntil (NVA)
        print ''Woken␣up␣after␣sleeping!''
end
```

Broadcast RTS To keep the illustration simple, we have provided pseudocode for only one role at each node. The nesC pseudocode is given below.

Theme: Simulation of RTS/CTS Handshake In this chapter, we have illustrated several protocols and shown how they can be coded in the nesC language. We observed that coding the protocols themselves was fairly straightforward.

```
Node B:
      booted():
      send (RTS to C);
      receive():
      if (CTS is received)
         transmit data;
      if (ACK is received)
         do nothing;
      Node C
      booted( )
      {NIL}
      receive():
      if (RTS received)
            send (CTS);
      if (data received)
            store (data);
      if (end of data received)
            send (ACK);
      Node A
      booted():
      {NIL}
      receive():
      if (RTS received) then
            extract duration d1 from the packet;
      lowPowerSleep(for duration d1);
      //that is , set NAV = busy;
      Node D
      booted():
      {NIL}
      receive():
      if (CTS is received) then
            extract duration d2 from the packet;
      lowPowerSleep(for duration d2);
%     //that is, set NAV = busy;
```

PROBLEMS

8.1 How does the S-MAC differ from traditional wireless MAC?

8.2 What are the modes of operation of S-MAC?

8.3 What is *idle listening*?

8.4 What are the sources of energy drain in a sensor node?

8.5 What is synchronization? Show an implementation in which five nodes synchronize with each other.

8.6 Describe the major contents of a routing table and give an example of nodes in a network slowly building their routing neighborhood routing table.

8.7 Provide an example where data are routed using the previous routing table and describe how a routing table changes over time.

8.8 Describe the CSMA-CA mechanism.

8.9 Describe the TDMA mechanism and its advantages and disadvantages for sensor network applications.

REFERENCES

1. *Crossbow Reference Manual*, Courtesy Crossbow Inc.
2. J. Agre, L. Clare, and S. Sastry, A taxonomy for distributed real-time control systems, *Adv. Comput.* **49**:303–352 (1999).
3. K. Chakrabarty, and S. S. Iyengar, *Scalable Infrastructure for Distributed Sensor Networks*, Springer-Verlag, 2005.
4. J. Hill, R. Szewczyk, A. Woo, S. Hollar, D. Culler, and K. Pister, System architecture directions for networked sensors, in *In Architectural Support for Programming Languages and Operating Systems,* 2000, pp. 93–104.
5. S. S. Iyengar, R. L. Kayshyap, and R. N. Madan, Distributed sensor networks, *IEEE Trans. Syst. Man, and Cybernet.* **21**(5):1027–1031 (1991).
6. V. Iyer, S. S. Iyengar, N. Balakrishnan, V. Phoha, and M. B. Srinivas, Farms: Fusionable ambient renewable MACs, *Proc. Sensor Application Symp.* Feb. 2009.
7. A. Moitra, and S. S. Iyengar, Parallel algorithms for some computational problems, *Adv. Comput.* **26**:93–153 (1987).
8. J. Polastre, J. Hill, D. Culler, Versatile low power media acess for wireless sensor networks, *Proc. 2nd Int. Conf. Embedded Networked Sensor Systems, SenSys '04*, ACM, New York, 2004, pp. 95–107.
9. I. Rhee, A. Warrier, M. Aia, J. Min, and M. L. Sichitiu, Z-mac: *A Hybrid MAC for Wireless Sensor Networks*, IEEE Press, Piscataway, NJ, 2008, vol. 16, pp. 511–524.
10. C. Xavier, and S. S. Iyengar, *Introduction to Parallel Algorithms*, Wiley, 1998.
11. W. Ye, F. Silva, and J. Heidemann, Ultra-low duty cycle MAC with scheduled channel polling, *Proc. 4th Int. Conf. Embedded Networked Sensor Systems, SenSys '06,* ACM, New York, 2006, pp. 321–334.

9 Techniques for Protocol Programming

The function of good software is to make the complex appear to be simple.

—Grady Booch

Wireless sensor networks are made up primarily of spatially distributed autonomous devices that use sensors to cooperatively monitor certain physical entitle such as temperature, sound, vibration, and pressure at different locations. Most sensor nodes are used in one of two contexts: as distributed databases or for event detection [2]. Irrespective of the roles employed in a sensor network, communication plays a critical role in determining life span, routing speed, and ultimately the nature of data that can be communicated in wireless sensor networks. It is for this main reason that we discuss some of the available protocols and features supported by most MAC implementations [4,1].

The S-MAC is a medium access control protocol for wireless sensor networks. It has been optimized for wireless sensor networks offering some advantages of 802.11-based MAC implementations such as reduced energy consumption and support for self-configuration. In its development, three sources of energy waste were identified and improved, resulting in energy savings of 2–6 times over those of traditional 802.11-like MACs. Two of the three areas of improvement are discussed below:

- *Collision.* When two or more nodes try to transmit packets at the same time, collisions may occur, resulting in packet corruption. Over a period of time these collisions represent a significant source of energy drain due to the retransmissions necessary after packet loss.
- *Overhearing.* In the case of over hearing, nodes listening to a channel pick up packets not intended for themselves, thus consuming more energy. With the S-MAC implementation, neighboring nodes form virtual clusters and autosynchronize on sleeping schedules, thus reducing occurences of overhearing.

Fundamentals of Sensor Network Programming: Applications and Technology, By S. S. Iyengar, N. Parameshwaran, V. V. Phoha, N. Balakrishnan, and C. D. Okoye

In the following sections we discuss the programming challenges involved in implementing a few basic protocols from the WSN [3] area.

9.1 THE MEDIATION DEVICE PROTOCOL

We now consider a well-known protocol called the *mediation device protocol*. In this protocol, node A wants to transmit a packet to node B:

1. Node A announces this to the mediation device by periodically sending a request-to-send (RTS) packets, which the mediation device captures. Node A sends its RTS packets instead of its query beacons (and thus they have the same period).
2. Node A then goes to receive mode (listens).
3. The mediation device waits for B's query, and replies to B with a query response packet, indicating A's address and a timing offset.
4. Node B now knows when A will be listening again, and sends a CTS to A. (Now, B also knows A's period, and thus knows at what time t' A's transmit mode will occur again.)
5. Node B waits at t' to receive data from A.
6. At t'' B will send its acknowledgment packet. (B knows t'' since it knows A's period.)
7. After the transaction has finished, A restores its periodic wakeup cycle and starts to emit query beacons again.
8. Node B also restores its own periodic cycle and thus decouples from A's period.

In the simple protocol shown above, we have assigned a single role to each node. Under each synchronous operation we have used a simple collection of rules that fire at appropriate instances. We use a periodic clock to send the RTS and to send the data. When a reply is received, we set a flag that selects which send operation should be executed next when the command `fired()` is executed. The details in pseudocode is given below.

```
Node A:

begin:

         Sensor.boot()
         StartTimer (Timer0,10 khz)

         //FLAGS
         sendRTS ↤ 1
         receiveRTS ↤ 0
```

```
        sendData ↤ 0
        receiveACK ↤ 0
        receiveCTS ↤ 0
        if (Timer.Fired() equals 'TRUE'):
                if (sendRTS equals 1):
                        send (RTS, MediationDevice)
                endIf
                setLowPowerListeningCycle(10%)
                receiveCTS ↤ 1

                if (sendData equals 1):
                        send (Data)
                        sendData ↤ 0
                        receiveACK = 1
                endIf
        endIf

        if (Receive.receivedData( ) equals 'TRUE'):
                if (receiveRTS equals 1):
                        if (receiveCTS equals 1):
                                receiveRTS ↤ 0
                                sendRTS ↤ 0
                                sendData ↤ 1
                        endIf
                endIf
                if (receiveACK equals 1):
                        sendRTS ↤ 1
                endIf
        endIf
end

Node B:

begin:
        Sensor.boot()
        query(MediationDevice)

        //FLAGS
        receiveResponse ↤ 1
        queryResponse ↤ FALSE

        receive.receivedData( ):
                if (queryResponse equals 'TRUE'):
                        time1 ↤ extractTime (CTS_PACKET)
                        wait( )
```

```
                        send (CTS_PACKET)
                        receiveResponse ↤ 0
                        receiveData ↤ 1
                endIf

                if (receiveData equals 1):
                        processData()
                        time2 ↤ extractTime (ACK_PACKET)
                        wait( )
                        send (ACK_PACKET)
                        receiveData ↤ 0
                endIf
end
```

9.2 CONTENTION-BASED PROTOCOLS

Contention-based protocols refer to a class of communication protocols that govern how multiple transmitters (devices) can make use of the same channel while avoiding collisions that may occur. This is achieved by specifying a set of rules by which each transmitting device must abide.

9.2.1 Carrier Sense Multiple Access Protocol

The carrier sense multiple access (CSMA) protocol belongs to the class of contention-based protocols in which, before any information is transmitted, a transmitting node verifies that no other concurrent transmissions are taking place on the shared medium. The operation of a variant of the CSMA protocol with collision detection is illustrated by the state diagram in Fig. 9.1.

The pseudocode accompanying Fig. 9.1 is provided below.

```
begin:
        // FLAGS
        idle ↤ 1
        randomDelay ↤ 0
        Listen ↤ 0
        AwaitCTS ↤ 0
        AwaitACK ↤ 0
        Backoff ↤ 0

        if (idle equals 1):
                numtrials ↤ 0
                randomDelay ↤ 1
                idle ↤ 0
        endIf
```

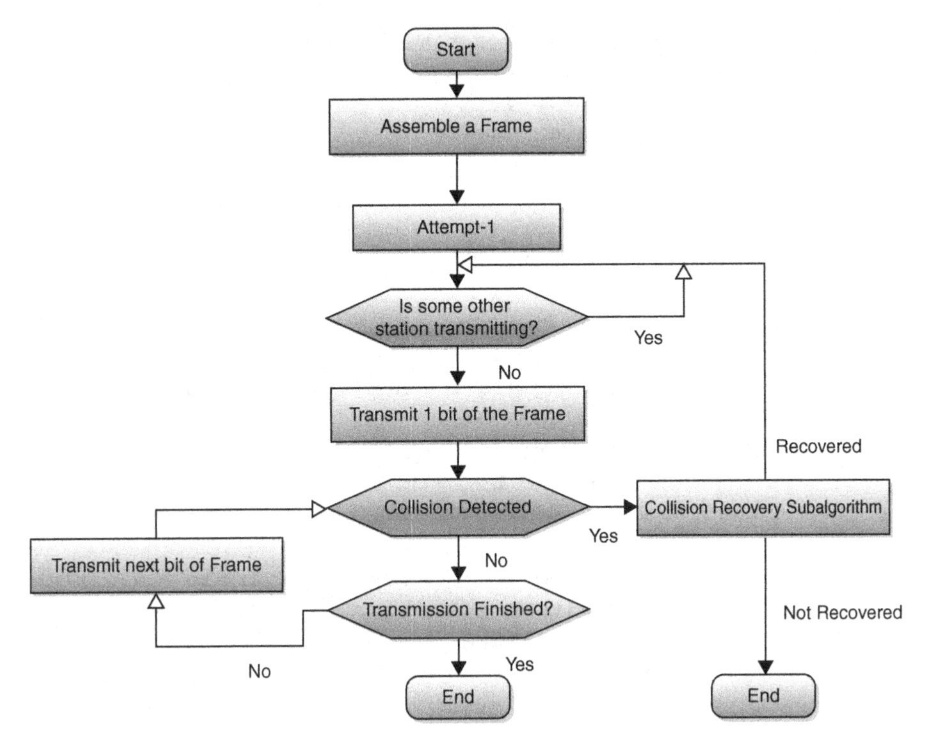

FIGURE 9.1 Carrier sense multiple access protocol.

```
if (randomDelay equals 1):
        Listen ↤ 1
        randomDelay ↤ 0
endIf

if (Listen equals 1 & mediumBusy equals 'True' &
numTrials equals... maxTrials):
        Listen ↤ 0
        Failure ↤ 'TRUE'
        idle ↤ 1
endIf

if (Listen equals 1 & mediumBusy equals 'TRUE' &
numTrials < maxTrials):
        Listen ↤ 0
        numTrials++
        Backoff ↤ 1
        setTimer( )
endIf
```

```
if (Listen equals 1 & idle equals 1):
        send (RTS)
        AwaitCTS ↤ 1
        Listen equals 0
endIf

if (BackOff equals 1 & timeout equals 1):
        BackOff ↤ 0
        Listen ↤ 1
endIf

if (AwaitCTS equals 1 & numTrials equals maxTrials):
        AwaitCTS ↤ 0
        Failure ↤ 'TRUE'
        idle ↤ 1
endIf

if (AwaitCTS equals 1 & numTrials equals MaxTrials):
        AwaitCTS ↤ 0
        numTrials++
        setTimer( )
        Backoff ↤ 1
endIf

if (AwaitCTS equals 1 & got (CTS) equals 'TRUE'):
        AwaitCTS ↤ 0
        send (Data)
        AwaitACK ↤ 1
endIf

if (AwaitACK equals 1 & acknowledgement()
equals 'FALSE' &...numTrials equals maxTrials):
        AwaitACK ↤ 0
        Failure ↤ 'TRUE'
        idle ↤ 1
endIf

if (AwaitACK equals 1 & acknowledgement()
equals 'FALSE' &...numTrials < maxTrials):
        numTrials++
        setTimer( )
        BackOff ↤ 1
endIf
```

```
        if (AwaitACK equals 1 & acknowledgement( )
        equals 'TRUE'):
                AwaitACK ↤ 0
                Success ↤ 'TRUE'
                idle ↤ 1
        endIf
end
```

9.3 PROGRAMMING WITH LINK-LAYER PROTOCOLS

In this section, we look at the programming challenges posed by the protocols at the link layer, and we start with the ARQ (automatic repeat request) technique. These protocols strive to send a packet more reliably using acknowledgments and resending.

9.4 AUTOMATIC REPEAT REQUEST (ARQ) PROTOCOL

The basic idea of ARQ protocols can be described as follows. The transmitting node's link-layer protocol accepts a data packet, creates a new packet by adding to it a header and a check sum, and transmits this packet to the receiver. The receiver verifies the checksum and accepts the packet if the checksum verification was successful, and sends a positive acknowledgment to the sender. If the check sum verification is not successful, a negative acknowledgment is sent. The transmitter, on receiving the positive acknowledgment, will know that the message was received successfully. If, on the other hand, the sender received a negative acknowledgment, it resends the packet.

9.5 TRANSMITTER ROLE

Note that the sender times out and exits to go to sleep if the negative acknowledgment is not received within a reasonable time. Furthermore, the sender keeps sending the message indefinitely as long as the receiver keeps sending it negative acknowledgments.

```
{
//Let p be a data packet coming from the MAC layer.

frame = Header++data packet p++check sequence (FCS);
// Construct the link layer packet

repeat forever
```

```
{
                transmit (frame, to receiver node j);
                // Can be done using Receive.receive( ).
                parallel
               {
        P1: {wait-for (ack);
        if (ack == positive) exit repeat loop;
        else
         continue; // it is negative ack
                      // waited long enough.
        P2: {if time-out() exit repeat loop; }
                } // end of parallel
        } // end of repeat;
        sleep( ) until woken-up( );
}

{
Receiver
node
j
    {
repeat for ever
    {
               // Can be done using Receive.receive ( ).
    p = receive-packet( );
    result = checksum-test (p);
    if (result == success)
        {send ("success"); exit repeat loop;}
    else send ("failure");
    } // end of repeat loop
    sleep ( ) until woken-up( );
}
}
```

The receiver node sends a positive acknowledgment and goes to sleep if it receives uncorrupted data. Otherwise, the receiver, after sending a negative acknowledgment, waits for a retransmission of the data.

```
{
TRANSMITTER(ARQ)

FSM Model
    Transmitter:

s0: frame = getFrame ( );
```

```
transmit ( frame );
go to s1;

s1:
receive( );
if acktype = +ve then {frame = getFrame( );
     transmit (frame)}
   else
     { transmit (frame);}
   upon ack go to s1;
   // Use Receive.receive( ) for this.
----------------------
global: frame, acktype = +ve;
booted( ):
   frame = getFrame( ); transmit (frame)

receive( ):
   actype = extract_ack type( );
   if acktype = + ve then {frame = getFrame( );
   transmit (frame)}
   else
      { transmit (frame); }
}

{
         RECEIVER
         fired( ):
         {NULL}

         receive( ):
  if checksum is valid then
  send +ve ack else send -ve ack;
}
```

If the negative acknowledgment was lost and the sender had timeout, then the receiver will end up waiting forever executing the `receive-packet()` statement. (See code for receiver node *j* above.) Excessive waiting results in battery power wastage.

9.6 ALTERNATING-BIT-BASED ARQ PROTOCOLS

A slight variation of the protocol above is the alternating-bit-based protocol, which attempts to transmit a packet to its neighbor. In this protocol, the transmitter uses a bit, called the *control bit*, which is set to 0 and 1 alternately. To start with, in order to transmit a packet p_1, the transmitter transmits prepends a 0 to P_1, producing a new packet $0{:}p_1$ and then transmits it to the receiver, sets it s timer, and waits for an acknowledgment (ACK). If timeout occurs before receipt of an ACK, the transmitter resends $0{:}p_1$, so that the receiver knows that the same data are being retransmitted. When the ACK is received from the receiver, the transmitter transmits the next packet p_2 prepending it with a control bit as $1{:}p_2$. If this is acknowledged, the transmitter transmits $0{:}p_3$,and so on. Thus, this protocol manages to transmit the data even when the ACK is lost.

```
{
Transmitter i Control bit b = 0;
REPEAT: repeat for ever
{
             read-packet-to-send (p);
     RESEND: transmit (b:p);
               PARALLEL:
             P1: { wait-for-ack( );
             p = p+1 mod 2;}
             // Receive.receive( );
             no clock necessary .
       P2: { time-out( );
       goto RESEND;
       }
}
```

```
{
         Receiver j
         Control bit b = 0;
         repeat for ever
         {
                  receive-packet (p);
                  // Receive.receive( );
                  is adequate to handle this situation.
                  b1 = extract-control-bit (p);
                  d = extract-data (p);
                  if (b1 = b) && checksum-valid (p) then
                  {send-ack( ); b = b+1 mod 2;};
                  // No -ve ack is sent.
                  // Go back and wait for the
                  // next transmission of the same data.
 }
}
```

```
{
         TRANSMITTER ⊣ alternating bit
         FSM Model
         Transmitter
         s0: bit =1; c bit = 1;
         frame = getFrame( );
         transmit (frame);
         // c bit is control bit (sent by receiver)
         go to s1;

s1: receive( );
       if c bit == bit
         { bit = bit + 1 mod 2; frame = bit: packet };
         transmit (frame);   upon ack: goto s1;
-------------------------
         Pseudo code for nesC:
         global: bit = 1, c bit = 1;
         booted( ):
  frame = getFrame( ); transmit (frame)

receive( ):
  actype = extract_acktype( );
  if acktype = + ve then
         {frame = getFrame( ); transmit (frame)}
  else  { transmit (frame);}
}


{
         Receiver ⊣ alternating bit
s0: control bit = 0; goto s1;

s1: receive( );
   b1 = extract-control-bit (p);
   d = extract-data (p);
   if (b1 = b) & checksum-valid (p) then
          {send-ack( ); b = b+1 mod 2;};
   // No -ve ack is sent.
}
```

9.6.1 A Generalized Version of the Previous Protocol

We can generalize the above protocol presented above by going back by n transmission steps and starting to retransmit them. We will illustrate this with an example below.

9.6.2 Example

Let $N = e4$. Let us use a buffer to store the packets. To start with, fill the buffer with eight packets, prepending each packet with a control byte, so that the content of the buffer buff will be buff = $0{:}p_0,1{:}p_1,2p_2,3p_3,4p_3,5p_4,6p_5,7p_6$. Start sending each packet from the left onward: $0{:}p_0,1{:}p_1,2{:}p_2,\ldots$ (one at a time) while simultaneously receiving the acknowledgments and processing them. Let b be the control byte of the last packet that was sent. The recipient is assumed to send a positive acknowledgment (+ACK) whenever the packet has been received correctly. Negative acknowledgments (indicating that the packet was not received successfully) are not sent by the recipient.

When the sender receives a +ACK, it first extracts the control byte, say, 0, from +ACK. It then can conclude that the packet $0{:}p_0$ has been successfully received. Suppose that it next receives a +ACK for which the control byte is 4. It then concludes that the packets $1{:}p_1$, $2{:}p_2$, and $3{:}p_3$ have not been received correctly. It thus sets the value of b to 1, and the packets are sent starting from the new value of b, rather than from it sold value.

The code below shows the generalized version of the ARQ protocol. Because of the multiple threads of reasoning involved, the control flow in the program is somewhat involved.

```
{
transmitter
A
{
   // Let Nmax = 4.
  repeat forever{
   load buffer [0:7] = 0: packet0: 1: packet1:.. .: 7: packet7;
   i = -1; // j is number of packets for which acks received.
   go = 1;
   PAR
P1: while (i < 7 )
         {
    shared:
              if (go) i = i +1;
              send (i: packet);
              go = 1;
    if (i ==7) wait( );
  }
P2:
wait-for-ack( );
         // Receive.receive( ); should solve this problem.
         j = extract-control-digit (ack);
         delete j: packet_j from the buffer;
shared: { i = j +1; go = 0}

} // for ever;
}
```

```
{
receiver B
  while (1) {
  receive ( from node a, packet p); // Receive.receive( ).
  contorl-digit j = extract-control-digit (from packet p);
  send-ack (control-digit);
  data = extract-data (from packet p);
  process (data);
  }
}


{
Transmitter ⊣ GENERALIZED VERSION
Transmitter:
s0: c=0, k=0; goto s1;
s1: while (k < 7) { send a[ k ]; k ++; };

s2: receive( );
    cbit = extract_cbit( );
    if (cbit == c) then (c=c+1 mod 8)
    else k = c; go to s1;
nesC code
booted( )
  c=0; k =0;
  transmit( );

transmit( ):
  while (k < 7) {send a [ k ]; k ++; };

receive( ):
  cbit = extract_cbit( );
  if (cbit == c) then (c=c+1 mod 8) else k = c;
  call transmit( );
}


{
Receiver ⊣ GENERALIZED VERSION

s0:    receive ( from node a, packet p);
     contorl-digit j = extract-control-digit (from packet p);
     send-ack (control-digit);
     data = extract-data (from packet p);
   process (data);
}
```

9.7 SELECTIVE REPEAT/SELECTIVE REJECT

The following program is somewhat simpler than the preceding one. In this, both the transmitter and the receiver have buffers. The transmitter keeps sending its data packets, while the receiver continues to receive them even if they are out of sequence. The receivers ends +ACK for received packets, and −ACK for missing packets. The transmitter then resends those packets for which $-v_e$ ACKs were received. While this protocol is more efficient than the ones above, the buffers that it uses will consume more memory in the sensor nodes.

```
Transmitter A
{
        buferl [0 .. 7 ] = 0: p0 ++ 1: p1 ++ 2:
                              p2 ++ ... ++ 7: p7;
        send-now [ 0 .. 7 ] = 0, 1 ,.., 7;
        send-next [ 0 .. 7 ] = non empty junk;

        j = -1; end =0;
        // initially start sending from bufferl [ ].
        while (send-next [ ] != empty)
        {
  PARALLEL:
  {
  PAR1:{
        for ( i = 0, i < size; i ++)
           {
           i1 = send-now[i];
           send (to node B, bufferl [i1]); }
           end = 1;
      }
  PAR2:{
        while (end != 1)
             {
              wait-for (ack, from node B);
              packet-id pid =
                   extract-control-byte (ack);
              ack-type = extract-ack-type (ack);
            if (ack-type = negative)
               { send-next [++ j] = pid; }
            } // end while
            send-now [ ] = send-next [ ];
                     size = size of send-next [ ];
       } // end PAR2
  } // end PARALLEL.
} // end while.
}
```

```
{
        receiver B
        buffer [ 0 .. 7 ] = empty;
        i = 0;
        while (i < 7)
        {
        receive (packet p, from node A);
        packet-id = get-packet-id (packet p);
        data = get-data (packet p);
        if (packet-id == i)
         { buffer [ i ] = data; i = i + 1; send (+ve ack)}
        else
         send (-ve ack, i);
        }
}
```

```
{
        Transmitter-Selective
               Repeat/Selective Reject
        s0: k = 0;
        s1: while (k < 7) send (a [k]); goto s2;
        s2: receive( );
        c = control bit( );
        if ack-for (c) is -ve then send a [c];
}
```

```
{
        Receiver-Selective Repeat/Selective Reject
        s0: i =0;
        while (i < 7)
        {
          receive (packet p, from node A);
   packet-id = get-packet-id (packet p);
   data = get-data (packet p);
   if (packet-id == i)
   {
     buffer [i] = data;
     i = i +1;
     send (+ve ack)
   }
   else
     send (-ve ack, i);
  }
```

```
nesC code:

booted( ):
   i = 0;
receive( ):
}
```

In some applications in the real world, it is more convenient to refer to the nodes using their network wide unique address than according to their location or data. In the following paragraphs, we briefiy discuss the need for address management in wireless sensor networks.

9.8 NAMING AND ADDRESSING

A sensor network is fragile, and the sensors can malfunction at any time. Thus, the nodes in a WSN must have the capability to allocate unique addresses to each individual node, and manage these addresses when new situations arise. Consider, for example, a node which acts as a bridge between two subnetworks, this node fails for some reason, the two subnetworks will lose connectivity, and for a node on one subnet, many nodes on the other sub node may become unreachable. While address management is a difficult problem, we will discuss a few simple techniques and their programming issues.

9.9 DISTRIBUTED ASSIGNMENT OF NETWORKWIDE ADDRESSES

This is a simple technique for assigning addresses to each node, but the technique is expensive, involving too many send operations. In this technique, each node generates a random address, sets it as its own address, and broadcasts to its neighbors to check whether it "clashes with any of its neighbours". If anyone reports address collision, the node again tries with another random address, and broadcasts to its neighbors again. This continues until no clashes occur.

```
module algorithm2C
{
    uses interface Boot;
    uses interface Random;
    uses interface Leds;
    uses interface AMPacket;
    uses interface Packet;
    uses interface AMSend;
    uses interface ActiveMessageAddress as AMA;
    uses interface PacketAcknowledgements;
```

```
    uses interface Receive;
    uses interface SplitControl as RadioCont rol;
    uses interface Timer <TMilli> as Timer0;
}
implementation
{
    void generateAddress( );
    void broadcastAddress( );
    am_group_t my group = 1;
    am_addr_t  my_address;
    bool conflict;

    event void Boot.booted( )
    {
        call RadioControl.start( );
    }

    /**************************************************
    Compute global addresss for each specific node and
    test if the address is in use
    **************************************************/
    event void RadioCont rol.startDone (error_t err)
    {
        // First we generate the address of the mote
        generateAddress( );

        // The initial address broadcast
        call Timer0.startOneShot (6000);

        // At this point, we can set the generated address
        call AMA. setAddress (my_group, my_address);
    }

    event void AMSend.sendDone (message_t* msg, error_t err)
    {
        if (err == FAIL)
        {
               call Leds.led0On( );
        }
    }

    event void Timer0.fired()
    {
        message_t msg;
        am_addr_t* temp;
```

```
        temp = (am_addr_t*) call AMSend.getPayload (&msg);
        temp = &my_address;
        call AMSend.send (0 x ffff, &msg, sizeof (am_addr_t));
    }

    /*****************************************************
    A simple function that randomly generates an address
    for a mote
    *****************************************************/
    void generateAddress( )
    {
        call Leds.set (0);
        my_address = (31071334523 % (call Random.rand16( )));
        }
        void broadcastAddress( )
        {
               message_t msg;
    am_addr_t* temp;
    temp = (am_addr_t*) (call AMSend.getPayload (&msg));
    *temp = call AMA.amAddress( );
    call AMSend. send (0 x ffff,&msg, sizeof (am_addr_t));
        }
        async event void AMA.changed( )
        {
        }

        event void RadioControl.stopDone (error_t err)
        {
        }

        event message_t* Receive.receive (message_t *msg,
        void* payload,... uint8_t len)
        {
               am_addr_t broadcasted_addr;

               broadcasted_addr = (am_addr_t*) payload;
               call Leds.set (0); // Clear debug leds
               if (*broadcasted_addr ==
               (call AMA.amAddress( ))) // Generate a new
               // addr
               {
                      call Leds.led0On( );
                      // A collision has occured
                      generateAddress( );
                      call AMA.setAddress
```

```
                    (my_group, my_address);
                    broadcastAddress( );
                    // Only broadcast new address
                    // after a collision
            }
            else
            {
                    call Leds.led1On( ); // No collison yet
            }
            return msg;
        }
}
```

In implementing this scheme, we use, as shown in the program presented above, a set of interfaces to achieve booting of the component, random-number generation, sending and receiving messages with the neighbors, and a timer to initiate the first broadcasting of a randomly generated address. We have also shown above which implementation has been chosen for each interface. In particular, the interface RadioControl (i.e., SplitControl) has been implemented using the ActiveMessageC component.

```
#define MOTE_AM_ID 10

configuration algorithm2AppC { }

implementation
{
    components MainC;
    components RandomC;
    components LedsC;
    components new AMSenderC(MOTE_AM_ID);
    components new AMReceiverC(MOTE_AM ID);
    components ActiveMessageAddressC;
    components algorithm2C;
    components ActiveMessageC;
    components new TimerMilliC( );

    algorithm2C.Boot->MainC. Boot;
    algorithm2C.Random->RandomC;
    algorithm2C.Leds->LedsC;
    algorithm2C.AMPacket->AMSenderC;
    algorithm2C.Packet->AMSenderC;
    algorithm2C.AMSend->AMSenderC;
    algorithm2C.AMA->ActiveMessageAddressC;
    algorithm2C.PacketAcknowledgements->AMSenderC;
```

```
    algorithm2C.Receive->AMReceiverC;
    algorithm2C.RadioControl->ActiveMessageC;
    algorithm2C.Timer0->TimerMilliC;
}

void generateAddress( )
{
    call Leds.set (0);
    my address = (31071334523 % (c a l l Random. rand16( )));
}
void broadcastAddress( )
{
        message_t msg;
    am_addr_t* temp;
    temp = (am_addr_t*) (call AMSend. getPayload (&msg));
    *temp = call AMA. amAddress( );
    call AMSend. send (0 x ffff,&msg, sizeof (am_addr_t));
}
```

In addition to the commands and events from the interfaces, we also use two more local functions for generating random addresses, and broadcasting the addresses to neighboring nodes:

```
event void Boot.booted( )
{
        call RadioCont rol.start( );
}
event void RadioCont rol.startDone (error_t err)
{
    // puts the address in my_address.
    generateAddress( );

    // The initial address broadcast
    call Timer0.startOneShot (6000);

    // Check once if any one else has my_address.
    // If not, at this point, we can set the
    generated address
    call AMA.setAddress (my_group, my_address);
}
```

The variable *my address* is a global variable in the implementation section of the component where the randomly generated address is stored before being broadcast to the neighbors.

As soon as the component is booted, the component is "turned on" by `RadioControl.start()`, where we perform the following tasks. We first generate a random address, and then place it in the global variable *my address*, so that it can be used by other functions in the component. We then check whether any other node has the same address as ours (address collision), and this is done by initiating a one-shot (monoshot) timer that fires after some delay(6000 ms) in our program. If no such collision is observed, we set the generated address as our address, and we are done. Our remaining task is only to reply to others' queries regarding whether there are any address clashes.

```
event void Timer0.fired( )
{
    message_t msg;
    am_addr_t *temp;
    temp = (am_addr_t *) call AMSend. getPayload (&msg);

    // storing the address of my_address in temp.
    *temp = my_address;

    call AMSend.send (0 x ffff, &msg, sizeof (am_addr_t));
}
event void AMSend. sendDone (message_t *msg, error_t err)
{
    if(err == FAIL)
    {
        // If send fails, switch on red led
        call Leds.led0On( );
    }
}


event message_t *Receive.receive(message_t* msg, void*
payload,  uint8_t len)
{
    am_addr_t* broadcasted_addr;

    broadcasted_addr = (am_addr_t*)payload;
    call Leds.set (0); // Clear debug leds
    if(*broadcasted_addr == (call AMA.amAddress( )))
    // Generate a new addr
    {
        call Leds.led0On( ); // A collision has occured
        generateAddress( );
        call AMA.setAddress (my_group, my_address);
```

```
        broadcastAddress( ); // Only broadcast
          new address after// a collision
    }
    else
    {
        call Leds.led1On( ); // No collison yet
    }
    return msg;
}
```

The monoshot time pulse triggers the event `fired()`, where we send the packet to the neighboring nodes. This is done by constructing the payload and storing *my address* as part of the packet, and then sending the packet. After sending the packet, if there are any errors reported in the send operation, we turn an LED on. After this we make no further attempts to send our address.

Whenever we receive a message from any node, we extract the sender's address, and compare it with ours to see if an address collision has occurred. If no collision was detected, then we have managed to find an address for our selves and remember this fact by setting the flag. (From this point onward, we only need to respond to others' messages by sending our address, without changing our address. This is achieved by setting the flag address Found to TRUE.) However, if we noticed a collision, we generate another address, set it as our own address, and broadcast it to all neighbors, and then wait to receive their replies. This continues until we find an address for ourselves.

Before we complete our task, we need to add some more functions, shown below.

```
void generateAddress( )
{
    call Leds. set (0);
    my_address = (31071334523 \% (call Random. rand16( )));
}

    async event void AMA.changed( ) { }

    event void RadioControl.stopDone (error_t err) { }
```

We generate a function using a large random number to minimize address collisions, and quick and definite convergence. We also note that we have nothing to do when ever the events `AMA.changed()` and `RadioControl.stopDone(...)` are invoked. Note that whenever the command `AMA.changed(..)` is called, `AMA.changed(..)` event is signaled. Similarly, `RadioControl.stop()` signals `RadioControl.stopDone(..)`. [Where is `RadioControl.stop()` called?]

9.10 IMPROVED ALGORITHMS

We will now discuss a couple of solutions that address the problems that we mentioned above.

9.10.1 Perkin's Solution

We will illustrate this solution with an example in which we are given a simple network with only a few nodes labeled. To start with, no nodes have any addresses assigned to them. Consider node *A*, which wants an address for itself. It generates a pair of random numbers $< t, f >$, where t is the *temporary address*, and f is the *fixed address*, and floods the network with this pair. Each node will receive a copy of this pair. When, for example, node *B* receives this pair, it checks whether if it already has an address that is the same as the number f. Since node *B* does not have any address to itself as yet, it sends a negative ACK message to node *A* using the temporary address *t*. Similarly, other nodes also send negative ACK messages to *A* using *t*. Node *A* waits for a sufficiently long time to receive these ACK messages, and finally chooses f as its address. Similarly, node *B* generates a pair of random numbers $< t_1, f_1 >$ and floods the network. Now, node *A* will check whether f_1 is the same as its address 1; if it is, it sends a positive acknowledgment, otherwise it sends a negative acknowledgment. Other nodes send positive acknowledgments. If node *A* has sent a negative acknowledgment, then node *B*, on receiving this acknowledgment, will generate a new pair of random numbers $< t_1, f_1 >$ and flood the network. Ultimately, node *B* finds an address, for example, 5, for itself. This process continues until all nodes find unique addresses for themselves. Of course, one problem with this algorithm is that the nodes do not know how long each of them should wait for the acknowledgments. Further, the random-number collisions can further delay the convergence of the algorithm.

The pseudocode for this algorithm is given below. If we generate the addresses from large address spaces, the collisions will be minimal. We also assume that each node knows the total number of nodes present in the network.

```
  assert f as own address.:

{
Node P
    temp-address-pool T= { some addresses };
    fixed-address-pool F = {ome-addresses };
    have-address = NO; kmax = 10;

    While (! have-address or k < kmax)
    // Get address for self.
    {
        t = randomly-select-an-address-from (T);
        f = randomly-select-an-address-from (F);
```

```
        // Announce: If any one has f
        // as his address, message me at
        flood (t, f);
        // address t;
        wait (ack, delta-t);
        // for delta-t units of time.
        if (ack = "Yes␣I␣have !")
        k++; // and continue.
        else
        { my-address = f; have-address = YES;}
}
    // I have an address now.
    // My job from now on is to reply to
    // other's queries.
    for (i = 1; i < m; i ++)
    { // forward m times. How to choose m?
    receive-requests-from-other-nodes (packet p);
    t = extract-temporary-address-from (p);
    f = extract-fixed-address-from (p);
    if (f == my-address)
  send (to address t, my-address);
    }
}
-----------------------------------------------------------
{
    PERKINS SOLUTION
    Assume: n total number of nodes .

s0: generate the pair <t, f >;
    send to all neighbours; goto s1;
s1: receive( );
    receive all acks; if all are -ve,
assert f as own address .;
    if any is +ve,
        goto s0;
    if it is a request for verifying
            some other node's␣request,
    then
       extract <f, t> par t, and check if f
          it clashes with its own address;
          if it does send +ve ack to t;

            if it doesn;t send -ve ack to t;
        nesC code:
// total number of nodes;
```

```
// T [ ] array of address.
global: n, T [ ];

booted ( ):
    call generate ( );

generate ( ):
generate <t, f >; send to all neighbours;

receive( ):
    check if all acks are received
                        // and stay in this state.
    and if a l l are -ve,
                        assert f as own address.;
    if any is +ve,
        call generate( );
    if it is a request_for
        verifying some other node's_request,
    then
       extract <f, t> part,
          and check if fit clashes
              with its own address;
    if it does send +ve ack to t;
    if it doesn't send -ve ack to t;
}
```

9.11 CONTENT-BASED ADDRESSING

Often in applications, a node may be interested in knowing about events that other nodes may have detected. It is possible that more than one node may have detected the nodes' events and stored the data in their memory. We now discuss a simple protocol where a node can query regarding an event to an unknown node. The node that is interested in the event (called the *sink node*) generates a query, and this is passed on to all nodes in the network. The node that can produce an answer to the query (called the *source*) responds to this by generating data packets and forwards them to the sink. The intermediate node, in addition to forwarding the data to the sink node, also stores the data along with the query in its cache, which it use later for any similar queries from future sink nodes.

Consider, for example, the network shown in Fig. 9.2, where node j (green) has detected an event E, and node c (yellow) is interested in knowing about this event, but it does not know about node j. Node c then sends a query along with its address, which is passed on to other nodes in the network. A node such as B will attempt to answer this query, but in this case, since B has no knowledge of the event, it passes the query to its neighbors. Eventually, the query reaches node j, which generates an answer to

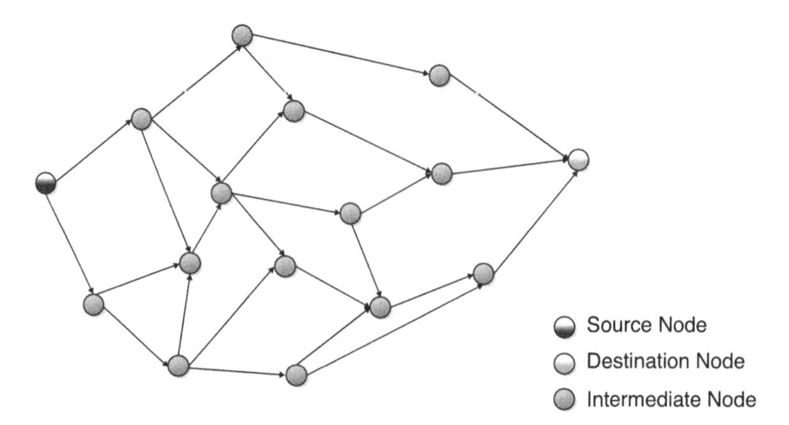

FIGURE 9.2 Flooding dataflow.

the query, and sends it to its neighbor, node k, for example, with a request to forward it to node c. Now, node k will store (i.e., cache) a copy of the reply, and forward it to one of its neighbors. This process continues until the reply reaches node c. If this query is initiated at any time in the future, the intermediate nodes that previously cached a copy of node j's reply in their local memory can generate a reply to the query initiator.

```
{
CONTENT BASED ADDRESSING self = initial Value();
Sink

-role:
    if self = = sink then
       flood network with request,
        establishing path to sink;

if self = = sink & reply received

    then accept it
    ;

source
role:
if self == source then send reply along path
;

intermediate
role:
if self
    == intermediate node on the path &
```

```
        have not already forwarded the
         request then forward the request
          and record the path
    ;
if self
    == intermediate node on the path &
        have not already forwarded the
            answer then forward the answer
                and store the answer
    ;
============================================================
    booted( ):
    self = initialValue( );
if self = = sink then
    flood network with request
     , establishing path to sink;
receive( )
:
if self == sink & reply received then accept it;
if self
    == intermediate node on the path &
        have not already forwarded the
            request then forward the request
                and record the path
    ;
if self
    == intermediate node on the path &
        have not already forwarded the
            answer then forward the
               answer and store the answer
    ;
}
```

We have thus far introduced several basic protocols and discussed their implementation details. We are now ready to design some real-world applications where we can use the techniques that we have learned so far, and we will be discussing some of them in the following chapter.

9.12 FLOODING

In WSN applications, sending data from one point (node) to all nodes in the network is one of the necessary basic behaviors. This happens in situations such as event monitoring, where a node wants to inform some other node (called the *sink node*) in the network about the event that it has detected but has no knowledge of. One solution to this problem is that node A can choose to flood the network with the data that it has at hand. In flooding, node A transmits its data to all its neighbors, with a request that each of its neighbors should forward the data to their neighbors with a similar request to forward. This continues until all the nodes in the network have received the data and have forwarded them with their requests to all their neighbors. Obviously, this way of passing data from a source node to a sink node is expensive, but sometimes this may be the only way to send data. In the following text, we first present the pseudocode for flooding, and then discuss the nesC programming challenges that we face while implementing it.

```
{
// Node A, the source node ⊣
// This node initiates the flooding.
{
        // from sensor or user.
        // Higher level behavior.
        sense-data (packet p);
    Data d = <packet p, command "forward">
    build-neighbourhood-table (table T);
    for each node k in the t able T
        // Or broadcast to every one?
        send-by-unicast (data d, to node k);
        sleep( ) until woken-up( );
}
// Node i, where i is any node except node A.
{

    build-neighborhood-table (T);
    wait-for (Data d);
    // Higher level behavior.
    for each node k in the table T
    // Or broadcast to every one?
    send-by-unicast (data d, to node k);
    sleep( ) until woken-up( );
}
}
```

In the pseudocode above, we have used two higher-level behaviors in the implementation of our solution to flooding: `sense-data(packetp)` and `sleep() untilwoken-up()`. Both these behaviors are implemented using the S-MAC schedule based listening and sleeping behaviors.

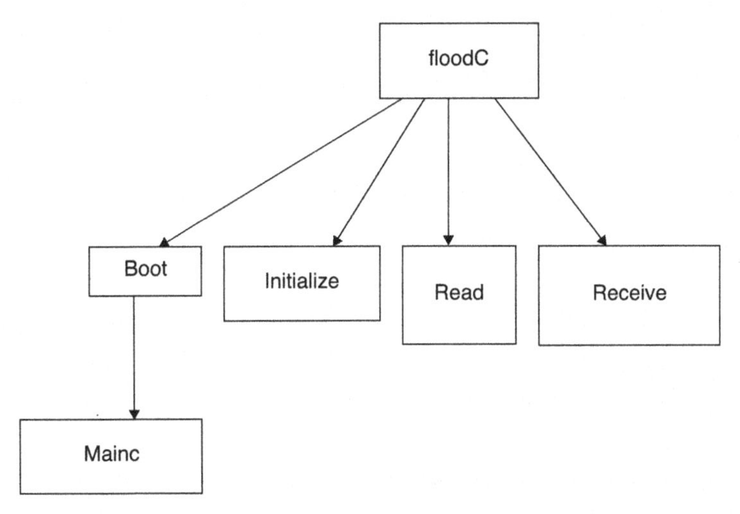

FIGURE 9.3 Flooding protocol architecture.

```
{
interface initialize;
command initialize( );

interface table:
    command build( );
event buildDone (err);

module floodC
    uses interfaces: initialize, read, flood, table;

booted( ):


    FLOODING
    Pseudo code

    source role:
    read data and flood the network;

sink role:
    reveive data;
process;

intermediate node role:
```

```
    if (received first time)
        flood the network;
------------------------------------------------------------
pseudo nesC Components:

    Assume: Let T be the neighborhood table;

booted( ):
    self = initialize role( );
firstTime = true;

if self = = source then read (x)
    ;
readDone( ):
    d = data from read operation;
flood the network;

receive( ):
    if self == sink then extract packet and process;

if self == intermediate & firstTime then
    {
    extract data;
    flood the network;
    firstTime = false;
    }
}
```

In this example, we illustrate how we write pseudocode specifying the behavior of the roles that each node plays during execution of this protocol from algorithmic description. We then convert this into another pseudocode, that is suitable for nesC programming. In the source role, the node reads the data and forwards them to every neighboring node. In the intermediate node role, any data received are forwarded to all neighboring nodes. In the sink node role, the data received are processed and stored for future use. Now, from the behavior observed in these individual roles, we can build the pseudocode for nesC programming. This basically involves distributing the role behaviors across the commands `booted()`, `readDone()`, and `receive()`, where a node chooses a role initially to guide the execution of the nesC (pseudo)code.

9.13 RUMOR ROUTING

In rumor routing, which is somewhat similar to flooding, a node *X* has an event *e* in which other sensors may be interested. In order to achieve this goal, *X* lets some of its neighbors *A* know about the event. (*Note:* In flooding, all neighbors are informed about the event.) Now node *K* forwards the data about the event to some of its neighbors, but at the same time records in its memory that *X* is its neighbor from where event *e* originated. This information is necessary if some node wants to reach *X* later. In the Fig 9.3, we thus see that *X* chooses to inform its neighbors *A* and *E* about the event, ignoring other neighbors. At this point, some of the nodes at hop distance 1 know about the event and that there is a route to the node *X*. The routing thus continues for some further hop distance, and terminates in this example at hop distance 4, where nodes *D* and *G* also know about the event at *X*.

9.13.1 Example

In the pseudocode below, each node chooses a few nodes randomly from its neighbors. We have assumed that the node *X* goes to sleep after transmitting its data, whereas the other nodes are still a wake, performing other activities.

Unlike in the scenario flooding, rumor routing is followed by the next phase where other nodes want to send queries to node *X* to learn more about the event. The program below shows what the node should do depending on what its role is. *A* node can play any of the three roles shown in the pseudocode presented below:

```
{
source role:

{
Program Code:

Node
X // Rumour initiator, wants to report an event e to
another node Z.

   {
       build-neighbourhood-table (T);
       for each node k from T
       {
       d = randomly decide to send ( );
       if (d == yes) send (to node k, own address, event e );
       } // end for loop
       sleep( ) until woken-up( );
       }
       Other intermediate nodes Y
       {
               (flag = 0); build-neighbourhood-table (T);
               if (flag == 1)
       display ("I was selected once before,
             and I forwarded the packet!");
```

```
                else
                {
                for each node k from T
                {
                d = randomly decide to send ( );
                if (d == yes)
                send-packet (to node k, own
                                address, event e);
                }
        flag = 1;
                }
        }
}
read data and perform rumour routing

sink role:
receive and process data

intermediate node role:
if ( received first time )
        flood the network;
}


{
Assume: Let T be the neighborhood table;
int self, first Time;
am_addr_t sensoraddress = 2;
// Pre-determined address of current sensor
am_group_t sensorgroup = 1;
RoutingData table [NO_OF_MOTES];
// Assume we already have this table.

booted( )
:
self = initialize Role( );
firstTime = true;
call RadioCont rol.start( );
event void RadioControl.startDone (error_t err)
    {
    // First we set the current address of this mote
    call AMA.setAddress (sensorgroup, sensoraddress);
    if self = = source then read (x)
        ;
    }
```

```
readDone( )

: // Note: This is executed only by source node.
    if self == source then
    {
    d = data from read operation;
    for each randomly selected neighbour k,
      send (to node k, data d);
}
event void AMSend.sendDone (message_t *msg, error_t err)
    {
    if (err == FAIL)
        {
        call Leds.led0On( );
        }
    }
receive( )
:
if self = = sink then e
xtract packet and process;

if self = = intermediate &firstTime then
    {
    firstTime = false;
    extract data;
    for each randomly selected neighbour k,
              send (to node k, data d);
    }

event void AMSend.sendDone (message_t* msg, error_t err)
    {
    if (err == FAIL)
        {
        call Leds.led0On( );
        }
    }

async event void AMA.changed( ) { }
event void RadioControl.stopDone (error_t err) { }
}

module Tracking
    {
    uses interface Boot;
    uses interface ReadStream as Temperature;
```

```
    uses interface Timer<TMilli> as Timer0;
    uses interface Timer<TMilli> as Timer1;
    }
```

9.14 TRACKING

In this section we show how as imple program-tracking utility can be written for a sensor network. In this example, it tracks the changes in temperature over time.

We now build the component for tracking by using the interfaces as shown.

```
{
configuration TrackingAppC
{
}

implementation
{
        components TrackingC as App;
        components MainC;
        components new TimerMilliC( ) as Timer0
        components new TelosbSensorC( )
        App.Boot ↦ MainC;
        App.Timer0 ↦ Timer0;
        App.ReadStream ↦ TelosbSensorC
}
}
```

`Read Stream` is needed to read the temperature from the environment, and timers are needed for sampling. We first read the initial values to store them in a vector. Later, during tracking, we sample and read the sensor values, and compare the new values with the values in the vector, and look for significant changes in the values. The implementations of interfaces are shown below.

We have chosen the `ReadStream` implementation for the TelosB mote. Using these implementations of the interfaces, we implement our component as follows. First, We start the two timers, one single shot and the other periodic. When the single shot fires, we sense the initial temperature and store it as the reference value. The periodic clock is used to sample the environment periodically and check whether the temperature has changed significantly.

```
{
implementation
{
    define #threshold 0.5
        int temp1, temp2;
```

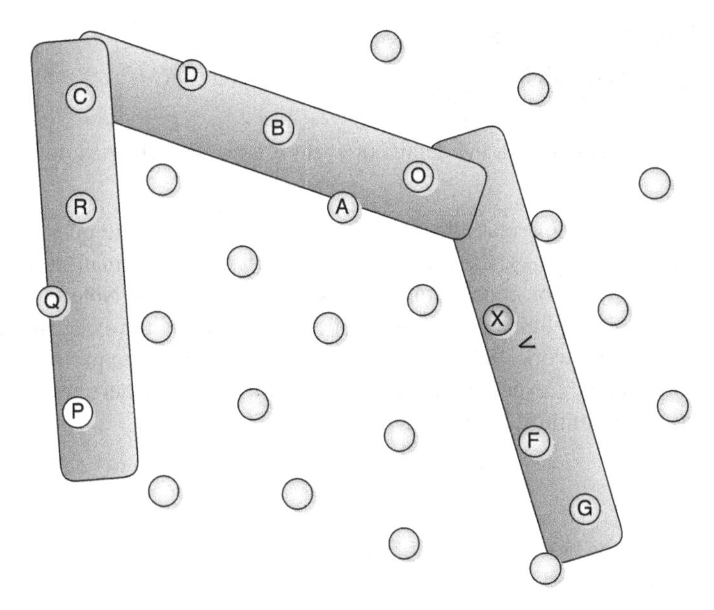

FIGURE 9.4 Querying in rumor routing.

```
        event void Boot.booted()
        {
    initialize two timers as
        Timer0.startOneShot (100) and
    Timer1.startPeriodic (2000);
 }
        event void Timer0.fired()
        {
    temp1 = Temperature.read( );
 }

event void Timer1.fired( )
        {
    temp2 = Temperature.read( );
     if (abs (temp1-temp2) > threshold)
        call led0Toggle( );
  }
}
```

9.15 QUERYING IN RUMOR ROUTING

For example, in Fig. 9.4, node *P* wants to know the event that occurred at node *X*, and sends a query. The sequence of events that take place are as follows:

1. Node P randomly chooses Q, and sends the query.
2. Since Q doesn't know the answer, it forwards it to R.
3. Node R forwards it to C.
4. Node C replies with the details of event e, as well as the path to the sensor X.

The program below classifies the nodes into four types: the querying node (P), the destination node (X), the node that has a path to the destination node (C), and the node that does not know about the destination node (Q). Note that this type of classification is implicit rather than explicit by virtue of the code that each node executes in the network. Each node executes its function in the protocol, and all the nodes cooperatively execute the overall protocol scheme, where we have used the following programming abstractions:

```
{
wait
-for( ), receive-from-neighbour( ),
        and send-request( ) .
Node P⊣originator of query to node X;

P does not know the id of X.
    // A node P wants to know about the event e.
    {
    Table T = build⊣neighbourhoodtable( );
    a = randomly⊣choose⊣a⊣neighbour (from T);
    packet p = query q++event e;
    send⊣request (own⊣address, to a, packet p);
    // programming abstraction
    wait

    -for (reply r, from node a);
    event E = extract (r);
    display (message m);
    }
}


{
wait

-for(), receive-from-neighbour( ),
                and send-request( ).
    Node P⊣originator of query to node X;
P does not know the id of X.
    // A node P wants to know about the event e.
```

```
    {
    Table T = build⊣neighbourhoodtable( );
    a = randomly⊣choose⊣a⊣neighbour (from T);
    packet p = query q++event e;
    send⊣(own⊣address, to a, packet p);
    // programming abstraction
    wait
    -for (reply r, from node a);
    event E = extract (r);
    display (message m);
    }

Node Q
⊣ a node that does not know anything about X,
so only forwards the request
to a random neighbour like
node R.
    {
    flag = 0;
    Table T = build⊣neighbourhoodtable( );

    if (flag == 1)
        display ("Already_forwarded");
    else
        {
        // programming abstraction
        receive⊣from⊣neighbour (query packet p);
        sender P = extract⊣id⊣of⊣sender (p);
        event E = extract⊣event (p);
        a = randomly⊣choose⊣a⊣
                neighbour (from T) such that
                the neighbour is not
                 the sender node P;
        // programming abstraction
        send⊣request (own⊣address, to a, event E);
        // wait for reply.// programming abstraction
        wait

        -for (packet p, from node a);
        send⊣reply (to P, packet p);
        flag = 1;
        }
    }

Node C
```

```
⊣ a node that knows about the
        event node X via a neighbour.

    {
    // I have not forwarded this message before.
    flag = 0;
    Table T = build⊣neighbourhoodtable( );

    if (flag == 1)
        display ("Already_forwarded");
    else
        {
        receive⊣from⊣neighbour (query packet p);
        sender P = extract⊣id⊣of⊣sender (p);
        // I already have a pointer to B.
        neighbour B = lookup⊣local⊣memory( );
        send (packet p, to node B);
        wait

        -for-reply (from node B, reply packet q);
        send⊣received⊣reply (to P, reply packet q);
        }
    }

Node D
// Destination node D that reports event e
    {
    receive⊣from⊣neighbour (query packet p);
    sender P = extract⊣id⊣of⊣sender (p);
    extract⊣original⊣
                sender (node S, from packet p);
    new packet p = sender S++event e
                send (packet p, to P);
    }
}
```

Each abstraction specifies an elementary behavior that we find useful in our sensor programming experience. All these abstractions will be implemented using the S-MAC schedule-based listening and sleeping.

Thus far we have been looking at examples where nodes have to transmit and receive data in a general setting. Nodes do not know about each other. However, in many real-world situations, nodes have partial knowledge about the each other. The following code illustrates one such technique where nodes exchange information more efficiently under certain assumptions.

The program above shows the code for each node involved the querying operations for rumor routing. For compactness, the roles performed by each of these nodes as specified in the code can be merged together and installed on each node.2

```
{
role - query generator (like P):
  fired( ):
  for each randomly generated neighbour q,
  {
      store q 's␣id;␣send␣(query␣to␣q);
␣␣}
sendDone(␣): {␣NIL␣}

receive(␣):
if␣packet␣received␣from␣one␣of␣the
neighbours previously␣randomly␣␣neighbour,
then␣extract␣the␣answer;␣halt.

role␣-␣out␣side␣rumor␣domain␣(like␣Q):
fired(␣):
NIL.

receive(␣):
if␣query␣received␣from␣a␣node␣p␣then
{
␣␣store␣address␣of␣p;
␣␣for␣each␣randomly␣generated␣neighbour␣q,
␣␣{␣store␣q'
      s id;
  send (query to q);
  }
}
if packet received from one of the
neighbours previously randomly chosen neighbour,
then extract the answer;
send the answer to node p;

role ⊣ inside rumor domain (like C):
fired( ):
NIL
receive( ):
    if query received from any node r then
    {
        obtain answer to query from the local memory;
        also obtain address of a node (like B)
```

```
        which has path to the event node like E;
        send <answer, address> to node r;
    }
}
```

PROBLEMS

9.1 In your own words, describe the role of the mediation device protocol.

9.2 Write a simple program to automatically assign random node identification numbers to sensor motes.

9.3 Explain the function of the alternating-bit-based ARQ protocol.

9.4 What is querying? Give an example of a real-world case.

9.5 What are the programming challenges at the link layer? Explain why.

9.6 Implement a 100 random uniform distribution of nodes that are assigned different addresses.

9.7 Describe content-based addressing. Explain with examples how and where this addressing scheme would be useful.

9.8 What is flooding? Describe its use and drawbacks in the context of sensor networks.

9.9 What are some of the improvements over fooding? Explain any two.

REFERENCES

1. V. Bharghavan, A. Demers, S. Shenker, and L. Zhang, Macaw: A media access protocol for wireless LANs, *Proc. ACM SIGCOMM 1994,* 1994.
2. C. G. Cassandras and S. Lafortune, *Introduction to Discrete Event Systems*, Kluwer Academic, Jan. 1999.
3. J. Hill, R. Szewczyk, A. Woo, S. Hollar, D. Culler, and K. Pister, System architecture directions for networked sensors, in *In Architectural Support for Programming Languages and Operating Systems,* 2000, pp. 93–104.
4. W. Ye, F. Silva, and J. Heidemann, Ultra-low duty cycle MAC with scheduled channel polling, *SenSys '06: Proc. 4th Int. Conf. Embedded Networked Sensor Systems,* ACM, New York, 2006, pp. 321–334.

PART IV
Real-World Scenarios

10 Sensor Deployment Abstraction

> Simplicity and elegance are unpopular because they require hard work and discipline to achieve and education to be appreciated.
>
> —Edsger Dijkstra

As we move into the future, the effects of Moore's law will progessively make the unit cost of sensor devices more and more negligible. Indoor and even outdoor environments with access to the existing power grid will be fertile environments for large number of low-impact sensors. Similarly motor vehicles, which already use large number of microprocessors in their existing systems, will be able to easily evolve to include sensors and wireless communication devices as standard hardware. The effect of this will be a potentially information-dense environment, with overwhelming amounts of data available from anywhere inhabited by humans. In the long term, the greatest cost and the challenge involved in building sensor networks will rest not in the hardware, but in the software. In order to deal with large numbers of nodes measuring large amounts of data, whole new paradigms in data aggregation and network architecture will be needed. Networks on this scale will resemble a living system more than a series of traditional computer networks, and our approach to them must vary accordingly. The deployment of sensors often necessitates an understanding of certain requirements such as sensor coverage, redundancy, and network architecture providing the essence of sensing physical space. Applications such as multiple target tracking, environmental monitoring, health monitoring, energy usage monitoring, or other general security monitoring require varying degrees of sensor coverage. Certain networking abstractions must be created to ensure easy management of sensors irrespective of the coverage type (dense Vs. less dense networks). In the following sections we discuss some of the common abstractions used to facilitate the deployment and management of sensors.

10.1 SENSOR NETWORK ABSTRACTION

Sensor networks should have the following properties:

- *Self-configuration*—formation of networks with no human intervention
- *Self-healing*—automatic deletion/addition of nodes without resetting the entire network

Fundamentals of Sensor Network Programming: Applications and Technology, By S. S. Iyengar, N. Parameshwaran, V. V. Phoha, N. Balakrishnan, and C. D. Okoye

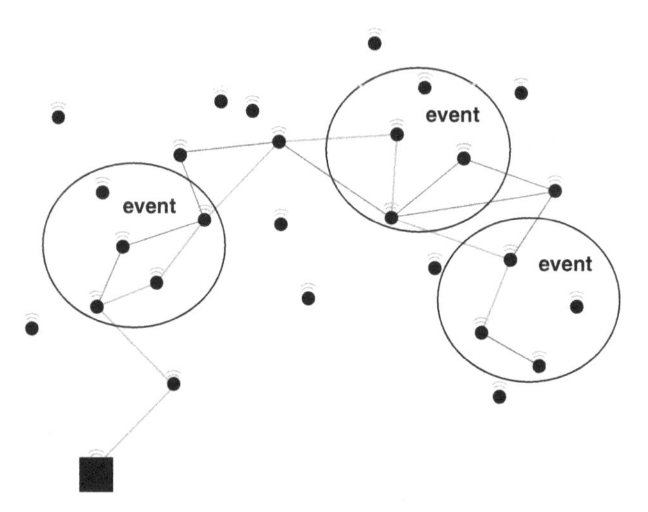

FIGURE 10.1 Data aggregation flow diagram.

- *Dynamic routing*—adapting routing schemes on the fly on the basis of network conditions such as link quality, hop count, and gradient
- *Multihop communication*—improving the scalability of the network by sending messages on a peer-to-peer basis to a base station

In order to support these functionalities at the large network level, programming design patterns and abstractions are essential. We will now present a few of them.

10.2 DATA AGGREGATION

In-network aggregation of data (see Fig. 10.1) is one of the fundamental data-processing techniques commonly used in sensor networks. This is because significant energy savings can be accrued over time as nodes collectively collaborate to forward data toward a data sink rather than having each node perform this task single-handedly. In TinyOS, some networking components having data aggregation capabilities have been provided such as the collection tree protocol and dissemination protocols.

10.2.1 TinyOS Data Aggregation Illustration

We present the following code to illustrate TinyOS data aggregation:

```
module AdvancedSendC
{
        uses interface Boot;
        uses interface Send as LeafSend;
        uses interface AMSend as SerialSend;
        uses interface StdControl as RoutingControl;
        uses interface SplitControl as RadioControl;
```

```
        uses interface SplitControl as SerialControl;
        uses interface Receive;
        uses interface Leds;
        uses interface Read<uint16_t>;
        uses interface RootControl;
        uses interface ActiveMessageAddress as AMA;

        uses interface Timer<TMilli> as Timer0;
        uses interface Queue<message_t*> as SerialQueue;
        uses interface Pool<message_t> as SerialPool;
}

implementation
{
        //Global Function Definitions
        void error (); //Called whenever an error occurs in our code
        void serialSuccess (); //Called whenever we send a serial packet
        // successfully
        void radioSuccess (); //Called whenever we send a radio packet
        // successfully
        task void sendSerialData (); //Actual task to send data to computer

        //Global Variables Declaration
        am_addr_t addr = 1;
        am_group_t group = 1;
        uint16_t periodic = 2000;
        uint16_t sensor_value;
        message_t buffer;
        bool serial_busy = FALSE;

        //Initialize some of our other components
        event void Boot.booted ()
        {
                call AMA.setAddress(group, addr);

                if(call RoutingControl.start()!= SUCCESS)
                        error ();

                if(call RadioControl.start() != SUCCESS)
                        error();

                if(call AMA.amAddress() == 1)
                {
                        call RootControl.setRoot(); // If your ID is 1,
                        // you are root
                        if(call SerialControl.start() != SUCCESS)
                                error();
                }
                else
                {
                        call Timer0.startPeriodic(periodic);
                }
        }
```

```
event void SerialControl.startDone(error_t err)
{}

event void RadioControl.startDone(error_t err)
{}

event void SerialControl.stopDone(error_t err)
{}

event void RadioControl.stopDone(error_t err)
{}

event void LeafSend.sendDone (message_t* msg, error_t err)
{}

async event void AMA.changed()
{}

event void Read.readDone(error_t err, uint16_t val)
{
        if(err != FAIL)
        {
                        sensor_value = val;
        }
        else
        {
                error();
        }
}

//Timer will only be fired on leaf nodes,
//When it happens, we send a reading.
event void Timer0 . fired ()
{
        uint16_t* data;
        if(call Read.read() == FAIL)
                error();
        data = (uint16_t*)call LeafSend.getPayload(&buffer);
        *data = sensor_value;

        if(call LeafSend.send(&buffer,sizeof(uint16_t))== SUCCESS)
                radioSuccess{};
}

//This receive method is only signalled on the root node. When it
//receives a packet, it tries to send it to the serial interface.
event message_t* Receive.receive(message_t* msg, void* payload,...
uint8_t len)
{
        message_t* serial packet;
        // First we check serial bus:
```

```
        if(serial_busy == FALSE)
        {
                serial_packet = call SerialPool.get();
                if(serial_packet == NULL) // Out of memory, so
                // drop all packets.
                {
                        error();
                        return msg;
                }
                else //Prepare packet for sending operation
                {
                        uint16_t* sense_value = (uint16_t*) call...
                          SerialSend.getPayload(serial_packet);
                        memcpy(sense_value, (uint16_t*) payload, len);

                        if (call SerialQueue.enqueue(serial_packet)...
                        == FAIL)
                        {
                                error();
                        }
                        else
                        {
                                serialSuccess();
                                serial_busy == TRUE;
                                post sendSerialData();
                        }
                }
        }
        else //If serial bus is busy we enqueue for now
        {
                serial_packet = call SerialPool.get();
                if(serial_packet == NULL) //Out of memory, drop
                //all packets
                {
                       error();
                       return msg;
                }
                else
                {
                       uint16_t* queue_value =(uint16_t*) call...
                         SerialSend.getPayload(serial_packet);
                       memcpy(queue_value, (uint16_t*) payload, len);

                       if (call SerialQueue.enqueue (serial_packet)...
                       == FAIL)
                               error();
                }

        }
return msg;
}
```

```
    event void SerialSend.sendDone (message_t* msg, error_t err)
    {
            serial_busy = FALSE;
            if (call SerialQueue.empty() == FALSE)
            {
                    serial_busy = TRUE;
                    post sendSerialData();
            }
    }

    task void sendSerialData()
    {
            buffer = *(call SerialQueue.dequeue());
            if(call SerialSend.send(0xffff,& buffer, sizeof(message_t)...
            == FAIL))
                    error();
            serial_busy = TRUE;
    }

    void error()
    {
            call Leds.led0Toggle();
    }
    void serialSuccess()
    {
            call Leds.led1Toggle();
    }
    void radioSuccess()
    {
            call Leds.led2Toggle();
    }
}
```

In this illustration of the collection tree protocol, every node in the network plays the role of either a leaf node or a root node. Roots advertise themselves throughout the network while leafs create routes to these roots using a routing gradient. The interface RootControl (we use in Section 10.2.1) provides the necessary functions to select roots and a special implementation of the AMSend interface allows leaf nodes to forward data to these root nodes.

10.3 COLLABORATION GROUP ABSTRACTIONS

We can view a collection of sensor nodes as a three-tuple $<S,R,M>$ where S is the set of nodes, R is a relation among the nodes, and M is the set of functions that the node performs. Each component of the tuple can be dynamically modified; that is, the set of nodes can vary as old nodes crash and new nodes are added, the relationship amongst them can be modified according to new situations, and new

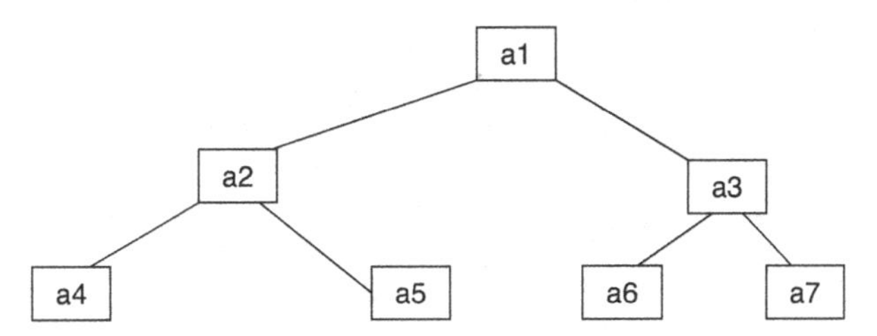

FIGURE 10.2 Typical operation of a node in this hierarchy.

functionalities may be added to the nodes. Abstract communities can be defined, and abstract routing protocols may be implemented using node level protocols. This also makes programming easier as the programmer at any time only needs to deal with abstract groups. By having more than one group of nodes perform the same set of functions, robustness may be improved.

10.3.1 Example

While the architecture of the group can be modeled as a directed acyclic graph, in this example we will assume that a hierarchical tree structure exists among the agents. Each node has a parent node (except the root node), and a set of child nodes (see Fig. 10.2). The overall task of a subgroup (formed by the nodes of a subtree) is broken down into subtasks by the node at the root of the subtree, distributed to the children, the results from the children are then aggregated, and finally sent to the parent.

```
// Typical operation of a node in this hierarchy.
// Given task T - e.g, measure temperature and brightness.
  while ( 1 )
        {
        decompose task T into subtasks T1,T2,...,Tn. //
          Assume n children.
        distribute subtasks to children (c1 ,..., cn);
        sense (x);
        r = receive results from children;
        q = compute (function f(x, r));
        communicate result q to parent;
}
```

10.3.2 Types of Group Abstractions

Abstractions using nodes can be defined based on geographical distribution of nodes, tasks assigned to each node and specific types of capabilities that may exist across the nodes. For example, in abstractions determined by geographically constrained group (GCG) of nodes, the nodes are placed within a chosen geographic area. The role chosen for each node will depend on relationship between the nodes and the

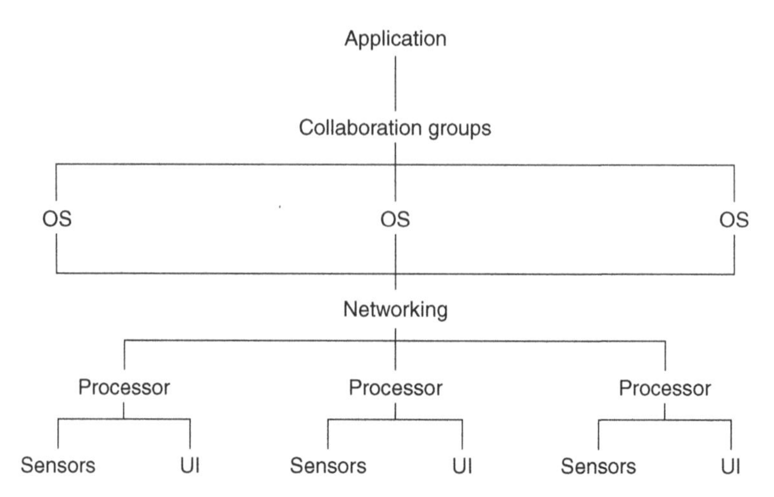

FIGURE 10.3 Nodes organized using geographically constrained group abstraction.

application. This abstraction will be useful when we want all the agents sensing the same environment. The exact relation between the nodes may be a tree or a directed acyclic graph.

10.3.3 Application of GCG Abstraction

Consider a task where we want to perform fusion of data collected over a specific geometric area. We can employ the GCG group abstraction as a solution to this problem [2,1]. In addition to low-level exceptions, there can be other types of exceptions such as communication holes arising in the geographic region. Various types of protocol supports can then be used to solve this problem. For example, one may choose to food all nodes in the entire geographic area (geocast strategy), route into the region and food (GEAR strategy; that is, geographic and energy-aware routing), food along the perimeter to route to a geographic location (GPSR strategy—greedy perimeter stateless routing), or food even with existence of holes (mobicast—mobile multicast).

10.3.4 *n*-Hop Neighborhood Group (*n*-HNG)

The *n*-HNG concept can be viewed as a notion capturing the idea of "reaching all nodes that can possibly be reached within a certain time." As before, the relation between the nodes may be hierarchical (a tree like structure). The advantage of this pattern is that it does not require the location of the nodes in order to be implemented. This pattern can also be used for local sensor selection. Exceptions can occur when communications fail, causing instability in hop counts. Solutions to this problem can include fooding, or ad hoc routing tree (explicit or implicit).

10.3.5 Publish/Subscribe Group (PSG) Pattern

This pattern is useful in situations where nodes of shared interests come together to perform a common task. For example, this works when all agents that provide certain data work together in data fusion and forward the results to an application. This pattern can be used in applications such as data/service discovery and pursuer/evader games. However, exceptional situations can arise when multiple nodes with shared interest food the network. Protocol support such as directed diffusion (which does not need geographic information), GHT (geographic hashing table: hashing data into geographic locations), and multirendezvous regions (replication of GHT in a region) may alleviate this problem.

10.3.6 Acquaintance Group (AG) Pattern

This pattern defines a group of nodes that "used to know each other." This captures the notion that "all nodes that share some state with each other" can fruitfully form a group. This is particularly suitable for nodes that are mobile. They may have peer-to-peer or leader–follower structure, and group membership can be historical or logical. The challenge in applying this pattern to real-word scenarios is to maintain connectivity among nodes, particularly when they are mobile.

With one or more fixed nodes and leader–follower structures, publish/subscribe protocols, and georouting (geographic routing) may be used. When all nodes are mobile and if the structure is peer-to-peer, structures such as Minimum Steiner Tree, Approximated Minimum Steiner Tree, or RoamHBA (roaming-hub-based architecture: maintaining a roaming backbone) may be used.

10.4 PROGRAMMING BEYOND INDIVIDUAL NODES

For large applications, we need to combine the techniques mentioned above and write programs building abstractions on top of the patterns of collaboration discussed above. This style of programming is not only scalable but also provides structural abstractions where each structure can be built, designed, implemented, and maintained separately (just as modular systems, or object-oriented systems).

PROBLEMS

10.1 List and explain some of the characteristics of wireless sensor networks.

10.2 Describe data aggregation and the concept of tree data structure.

10.3 Describe and list real-life examples of publish subscribe systems and how the mechanism can be adapted to sensor networks.

REFERENCES

1. V. Iyer, S. S. Iyengar, N. Balakrishnan, V. Phoha, and M. B. Srinivas, Farms: Fusionable ambient renewable MACs, *Proc. IEEE Sensors Applications Symp. SAS 2009,* Feb. 17–19, 2009, pp. 169–174.
2. M. B. Srinivas, V. Iyer, G. Rama Murthy, and B. Hochet, C-error simulator for development for sensor and location aware sensing applications, *Proc. 3rd Int. Conf. Sensing Technology*, Taichung, Taiwan, 2002, pp. 799–804.

11 Standards for Building Wireless Sensor Network Applications

Good design adds value faster than it adds cost.

—Thomas C. Gale

11.1 802.XX INDUSTRY FREQUENCY AND DATA RATES

The IEEE 802.15.4 is a standard that specifies the implementation details of medium access control (MAC) and the physicallayer (PHY) for low-rate wireless networks. It provides the fundamental lower network layers of a wireless network focusing on low cost and low speed typically associated with wireless sensor networks. Currently, it has been implemented by several networking solutions such as ZigBee, Wireless HART, and MiWi, which provide higher-level communication protocols in addition to the 802.15.4 standard. In this chapter, most of the focus will be on the ZigBee standard, due to its popularity among wireless networking vendors. A simple comparison of some wireless standards is shown in Fig. 11.1.

The ZigBee standard defines the security, networking, and application frameworks for an IEEE 802.15.4–based system. It creates a self-forming, self-healing mesh network capable of supporting thousands of wireless devices on a single network [1–3]. The ZigBee stack architecture is divided into five parts:

- Security service provider
- Application layer (support sublayer, framework, ZigBee device object)
- Network layer
- Datalink layer
- Physical layer

Most sensor applications based on ZigBee specifications typically interact with the application and network layers specified by ZigBee. The network layer provides necessary abstractions for managing multihop communication between the various nodes while the application support layer manages communication between the various application objects. Also, the ZigBee alliance makes provisions for security

Fundamentals of Sensor Network Programming: Applications and Technology, By S. S. Iyengar, N. Parameshwaran, V. V. Phoha, N. Balakrishnan, and C. D. Okoye

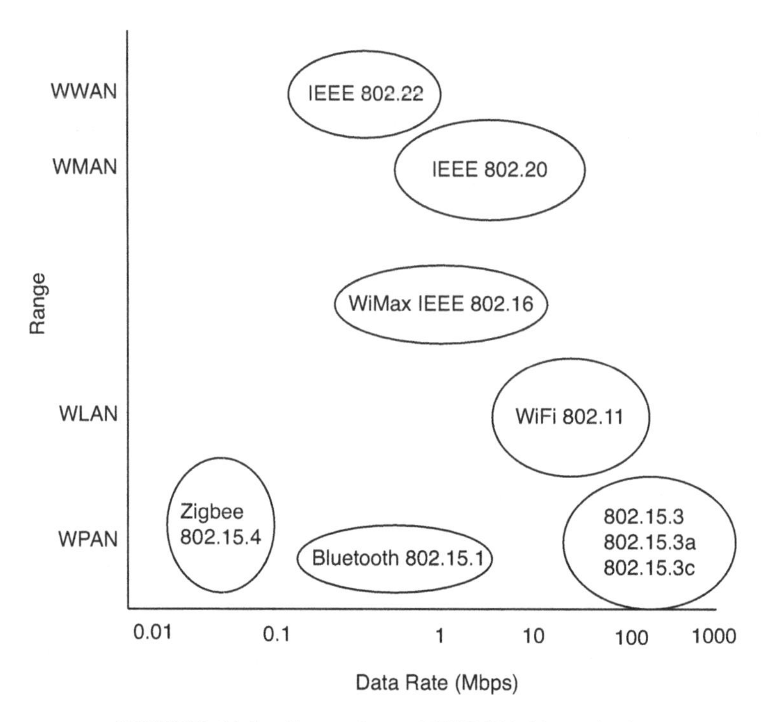

FIGURE 11.1 Comparison of IEEE 802.11 standards.

services, device management, and the ability to extend the application framework with vendor-specific services. The full architecture of the ZigBee stack is shown in Fig. 11.2.

11.2 ZigBee DEVICES AND COMPONENTS

The maintenance of the 802.15.4 sensor network is the backbone of the easily programmable model of the network stack design. As by design, they are self-organizing and easy to maintain using a hierarchical addressing scheme. ZigBee devices are divided into three major classes:

1. ZigBee network coordinator
2. ZigBee router
3. ZigBee end devices

Type 1 devices are typically addressed as node 0, acting as a communication gateway and control node. There is usually only one such node in a given cluster-based topology, and it can centrally address up to 2^{16} nodes, which have sufficient resources in terms of memory and data-handling capability. As it is the first full-function device

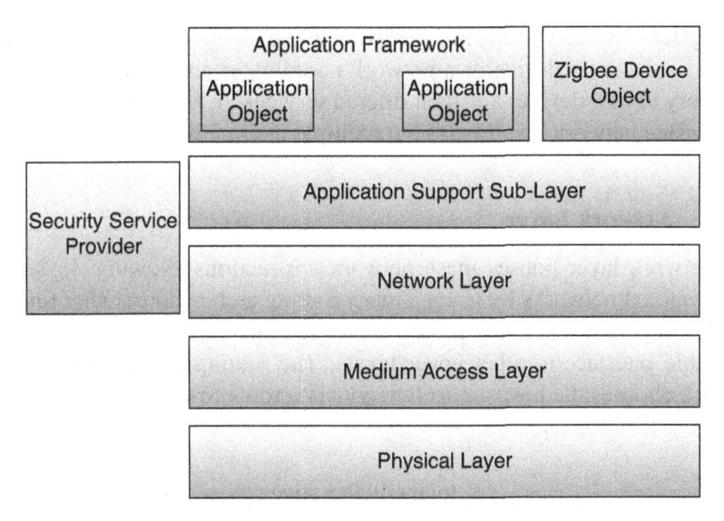

FIGURE 11.2 ZigBee architecture.

set up on the network, node 0 is responsible for topology discovery and initial assignment of addresses and identifying specific nodes as intermediate routers or end nodes to initialize the sensor network. Type 2 nodes are routers whose primary capability is to forward multihop communication from the data nodes to the central coordinator gateway. These nodes need to have a routing stack so that they may effectively participate in routing of the measured data from their next-hop neighbors. Type 3 nodes are the most adaptive as these are remotely placed on a long-term basis. They are equipped with physical sensors to accurately measure environmental values and transmit only when queried or when an event has triggered idling for most of its lifetime. The 802.15.4 stack code is designed as a state machine that allows a reentrant API library for sending nonblocking messages. The programming APIs cover each network layer individually such that it is documented as a library function; these APIs are compiled into native target processors with hardware-specific optimizations.

11.2.1 Application Layer

The application layer is defined by the ZigBee specification but is implemented by manufacturers. It consists of three critical components:

- Application support sublayer
- Applicationframework
- Application object
- ZigBee device object

The application support sublayer manages communication between the various application objects. Each application object provides specific sensing functionality for the ZigBee device. The ZigBee device object defines the role of each device in the network as an end device or a network coordinator and also is responsible for the discovery of new devices and their offered services. In any network, there can be only one ZigBee network coordinator but multiple intermediate routers and end devices.

11.2.2 Network Layer

The network layer houses mesh network abstractions allowing ZigBee devices to form resilient networks by using a mesh routing architecture. Other functions of the network layer are to enable proper use of the underlying MAC layer by providing a suitable interface to other upper layers. The routing protocol in use in a ZigBee network chooses the lowest-cost (energywise) route to the intended destination. The following factor is used:

$$\text{Global} = \frac{\text{total collaborative resource needed}}{\text{resource available in individual sensor}}$$

This factor is calculated on routing or completing a sensor network task without having a single failure, keeping cross-layer energy consumption to a minimum.

11.2.3 Datalink Layer

The low-level layers deal with radios and a suitable MAC that allows for communication with its neighbors without any central coordination. The datalink layer is also responsible for aggregating data from all its neighbors and sending an acknowledgment when it receives sufficiently reliable data to higher network layers.

11.2.4 Physical Layer

This layer consists of the actual radio and is IEEE 802.15.4–compliant.

11.3 ZigBee APPLICATION DEVELOPMENT

ZigBee basically uses digital radios to allow devices to communicate with one another. A typical ZigBee network consists of several types of devices. A network coordinator is a device that sets up the network, is aware of all the nodes within its network, and manages the information about each node as well as the information that is being transmitted/received within the network [2]. Every ZigBee network must contain a network coordinator. Other full-function devices (FFDs) may be found in the network, and these devices support all of the 802.15.4 functions. They can serve as network coordinators, network routers, or devices that interact with the physical world. The final device found in these networks is the reduced-function device (RFD), which usually serves only as a device that interacts with the physical world. An example

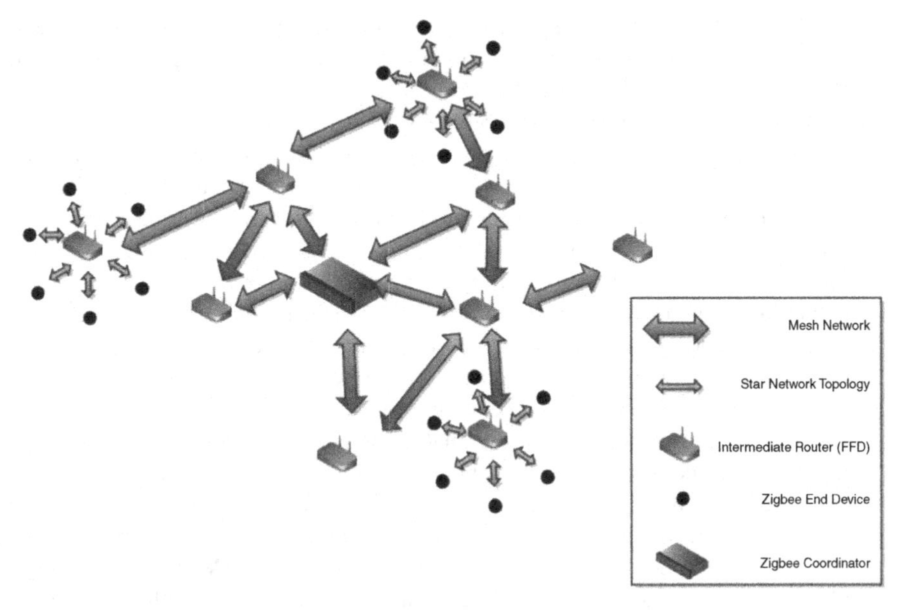

FIGURE 11.3 A typical ZigBee network.

of a ZigBee network is shown in Fig. 11.3. The components needed to develop a real-time automation application are

1. Coordinator node with network configuration and application
2. Routers that are in the form of clusterheads and allow multihopping
3. End devices connected to sensors

The ZigBee PAN coordinator which is currently the most powerful node on the network is normally 1-hop. Furthermore, the Wireless Sensor Network (WSN) communicates to a fixed Local Area Network (LAN) which has a gateway Internet Protocol (IP) address to transfer data serially or by built-in ethernet ports in some models of ZigBee.

Sometimes a Sniffer program from an IP can be used to monitor the unknown topology of a WSN network. Programming in the air can be used to correct some on-going problems, which may need reconfiguration through the use of low-level sensor programming. The coordinator node synchronizes all the member nodes and provides them the flexibility to communicate data in either in their assigned time-slots or by re-calculating a different slot. Moreover, the coordinator node acts as a router to route all data from its domain to coordinators in other domain (peer-to-peer structure) or to a higher-level coordinator (if it has a hierarchical topology).

The router nodes act like clusterhead and allow all their children to directly interface to enable routing. The main function of these router nodes is to provide reliable multihop paths in the event of a failure on the network.

The end devices are mostly sensors connected to physical devices and have very little memory requirements. They are mainly designed to be compatible with different manufacturers' specifications. This makes it possible to mix and match different sensors without any cross-compatibility issues.

The ZigBee Alliance provides a number of profiles that provide a framework as shown in Fig. 11.2 in which related applications can work simultaneously. In this way, end devices from different vendors can interoperate as long as they adhere to the given profile. One of these profiles is the "home control, lighting profile." This profile focuses on sensing and controlling light levels in the home environment. The profile defines different device descriptions that belong to the profile, including "light sensor monochromatic," "switch remote control," "switching load controller," and "dimmer remote control." A profile can consist of 216 device descriptors and can hold up to 256 clusters. Each cluster can contain up to 216 attributes. A device description contains a set of mandatory and optional input and output clusters from the profile. Input clusters consist of attributes that can be set by other devices; for example, the light sensor has an attribute called ReportTime, which controls the time interval between light readings. Output clusters consist of attributes that supply data to other devices; for instance, the "light sensor monochromatic" (LSM) has one attribute in its output cluster, named CurrentLevel, which holds the current light sensor reading measured in 6 lux. Mandatory clusters (including every attribute within these) must be implemented by the appropriate end devices. Optional clusters may be implemented, but if a device supports an optional cluster, it must implement every attribute within that cluster.

11.4 DISSEMINATION AND EVALUATION

As this research is a multidisciplinary effort, the programming of the sensor networks will be evaluated on the types of real-time and energy constraint needs, which have been addressed at the processing nodes and also how this stream of live data is been aggregated on the servers to archive and interface to standard IP-based networks or online queryable geographic information systems (GISs). Wireless ad hoc sensor networks (WASNs) have attracted much attention in recent years from a diverse set of research communities. Researchers have been concerned mostly with exploring applications scenarios, investing new routing and access control protocols, proposing new energy-saving algorithmic techniques, and developing hardware prototypes of sensor nodes.

PROBLEMS

11.1 Contrast the features of ZigBee standard with Bluetooth wireless protocol.

11.2 In three paragraphs (about 600 words), describe the application layer in ZigBee stack architecture.

11.3 What methods does ZigBee utilize for device discovery?

11.4 Describe the network layer in ZigBee stack architecture.

11.5 Describe the topologies supported by ZigBee.

11.6 Write short notes on the following ZigBee devices and components:

(a) ZigBee network coordinator

(b) ZigBee router

(c) ZigBee end devices

11.7 In about 500 words, explain the ZigBee "profile."

11.8 Write short notes (about 600 words) on the ZigBee cluster library, ZigBee binding, and ZigBee binding table.

11.9 Briefly describe the "lighting" profile in the ZigBee home control environment.

11.10 Research and prepare short notes on "switch" profile in the ZigBee home automation environment.

11.11 Figure 11.4 shows a room layout with two doors and four lights. Lights L_1, L_2, L_3, and L_4 are controlled by ZigBee enable switches S_1, S_2, S_3, and S_4, respectively.

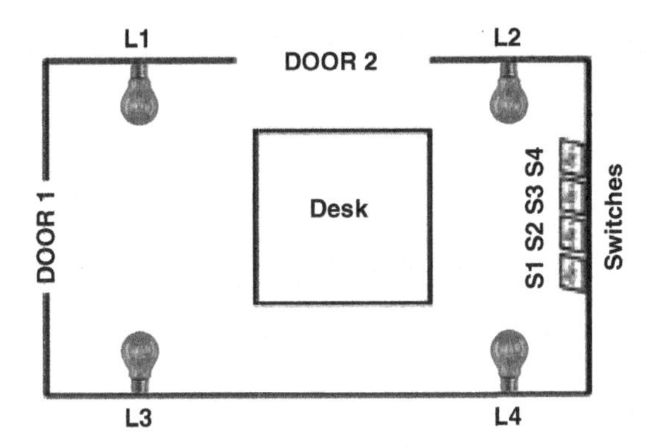

FIGURE 11.4 Problem 11.11

11.12 Design a ZigBee home automation application to perform the following tasks:

(a) Lights L_1 and L_4 should be switched ON when a person enters through door 1.

(b) Lights L_2 and L_4 should be switched ON when a person enters through door 2.

(c) Lights should continue to be ON when a person is present in the room.

(d) All lights should be switched OFF when a person leaves the room.

11.13 Discuss the equipment required (e.g., ZigBee enabled occupancy sensors, switches, controllers, gateways), and placement of the sensors, and explain in detail how you would implement the application.

11.14 Describe "Cyclic" and "Pin" sleep modes in ZigBee.

11.15 How do you create and maintain a list of active devices that are connected to a ZigBee network?

11.16 Discuss MessageFreshener timers in ZigBee.

11.17 How does ZigBee achieve low power consumption?

REFERENCES

1. ZigBee Alliance Overview, ZigBee Alliance.
2. ZigBee Wireless Networking, Newnes Publications, 2008.
3. D. Gislason, ZigBee Resource Guide, Webcom Communication Corp., 2008.

12 INSPIRE: Innovation in Sensor Programming Implementation for Real-time Environment*

The bottleneck in writing code isn't in the writing of the code, it's in understanding and conceptualising what needs to be done.

—Shane Legg

Wireless ad hoc sensor networks (WASNs) have drawn much attention in recent years from a diverse set of research communities. Researchers have been concerned mostly with exploring application scenarios, investing new routing and access control protocols, proposing new energy-saving algorithmic techniques, and developing hardware prototypes of sensor nodes.

This is described in the context of programmers and code base. Typically traditional programmers use lots of inefficient techniques, which need more computation. As in the context of event-driven programming it encapsulated as an even-handler. Making any unexpected loops to be present in the actual deployment.

Little attention has been devoted to actually measuring the lifetime of sensor networks. *Lifetime* is defined as the useful time during which the sensor networks provide live datastreams before sensor faults occur or a percentage of sensors completely cease to function. Since the normal sensor system's lifetime is in the order of many months to years, especially in the case of ultra-low-duty-cycling applications, it is difficult to deploy and predict the lifetime of applications. In this section we will use a software simulator that allows us to deploy a large sensor network. This will enable us to accurately measure lifetime in real-time clocks, sensor errors, and radio profiling during radio idle, receiving, and transmissions. As the practicality of the sensor network is based on limited power resources, we extend this power model by using renewable energy resources and its effects on the performance levels of different algorithms.

12.1 MOTIVATION AND BACKGROUND

A WASN is an ad hoc network of resource-limited, static, wireless, sensor nodes that are used to monitor dynamic physical processes. Typically a user queries the

* Portions of this chapter were contributed from various sources by V. Iyer [4,5].

Fundamentals of Sensor Network Programming: Applications and Technology, By S. S. Iyengar, N. Parameshwaran, V. V. Phoha, N. Balakrishnan, and C. D. Okoye

network, the query triggers some reaction from the network, and as a result the user receives the information needed. The reaction to the query can vary from a simple return of sensor value, to a complex unfolding of a distributed algorithm among some of all of the sensor nodes, such as a collaborative signal processing algorithm, a distributed estimation algorithm, or an actuator control algorithm. Furthermore, there are multiple users who are transiently connected to the network, each having different needs in requested information.

These systems are quite different from traditional networks: (1) they have severe energy, computation, storage, and bandwidth constraints; and (2) their overall usage scenario is quite different from that of traditional networks. There is not a mere exchange of two specific nodes. The user will be interested in some parameters of a dynamic physical process. To efficiently achieve this, the nodes have to form an application-specific distributed system to provide the user with the answers. Moreover, simulation results show that the local computation takes a fraction of the energy when compared to data transmission. So the design uses fixed resources such as a battery to do only sensing and local computations and uses a renewable energy resource to transmit data. The nodes that are involved in the process of providing data information are constantly changing as the physical phenomenon is changing. Therefore the user interacts with the system as a whole, which provides sufficient information without loosing its usefulness (drain due to power). The WASN is not there to be queried but instead provides information in an efficient, reliable, and collaborative mode wirelessely.

Advancements in wireless technologies have made it possible to integrate with radios that are capable of self-organizing and communicating in an ad hoc configuration with minimal infrastructure. Our system "INSPIRE" maintains a real-time network and periodically senses data from the onboard analog sensors, which collectively aggregate the sensed data by collaborative processing with other nodes. The technique employed is the divide-and-conquer technology in order to provide a robust network reliability for hazardous applications. In the following sections we provide a structured approach in terms of describing the new computational system INSPIRE in detail.

12.1.1 INSPIRE: An Introduction

Our system, INSPIRE (innovation in sensor programming implementation for real-time environments) adopts an active sensor approach to allow any distributed algorithm to be executed in the network. Energy is a critical resource in sensor networks. MAC protocols such as S-MAC and TMAC [6] coordinate sleep schedules to reduce energy consumption. More recently, low-power listening (LPL) approaches such as WiseMAC and B-MAC exploit very brief polling of channel activity combined with long preambles before each transmission, saving energy, particularly during low network utilization. Synchronization cost, either explicitly in scheduling, or implicitly in long preambles, limits all these protocols to duty cycles of 1–2%. We demonstrate ultralow duty cycles of 0.1% with the use of a mathematical model [5] for lifetime, and also support it with simulation results [4] that highlight the power savings due to overhearing and node density radio interference when idling. We use the extended work of a new MAC protocol called *scheduled channel polling* (SCP) [8]. This work prompts three new research directions that highlight the power savings for ultra-low-duty cycling and also concludes that the power consumption of SCP decreases with faster radios, but that of LPL increases. As in energy-harvesting applications during data forwarding, it may not be able to find forwarding nodes as they may be out of current radio range or might not be able to respond because the radio is disabled or battery is recharging. Therefore, we need to provide queuing of data messages carrying payloads, be able to build new neighborhood lists proactively according to the rechargeable time window, and be able to sequence the data values and timestamp from individual nodes so that the values can be aggregated as they arrive in the same sequence from source to sink. A service-based architecture is used that makes writing applications transparent to the data delivery mechanism in high-traffic or adaptive burst wireless traffic conditions.

12.1.2 RTOS Abstraction Layer

Whenever your application handles a variety of activities or manages multiple devices, a real-time operating system (RTOS) can help you simplify the code by separating different tasks. An RTOS is thus an important tool that allows you to "divide and conquer."

At its simplest level, an RTOS is just a context switcher plus a mechanism for handling some intertask synchronization (see RTOS module configuration in Table 12.1). The context switcher allocates the CPU to various tasks according to a scheduling algorithm. The tasks can therefore advance at different paces, depending on how many CPU cycles they obtain. An RTOS also provides mechanisms that allow the tasks to synchronize their activities. Examples of such mechanisms include various types of semaphores, mailboxes, and message queues. Although the basic services provided by RTOSs are very much the same, they are accessible through significantly different APIs. This poses a serious problem for many companies that want to deploy shared application code on different operating systems or that don't want to lock their strategic application into a particular operating system.

An active object-based framework such as the INSPIRE offers a more elegant solution.

12.1.3 Minimal Application

How software is implemented in sensor networks is essential to understanding sensor networks. A network architecture and protocols are essential foundations for building software applications. Active objects in microframework are encapsulated tasks (each embedding a state machine and an event queue) that communicate with one another

TABLE 12.1 Real-Time OS Module Configuration

Coresident OS	Tasks	Priority	Number of Control Points
Kernel	Control	Highest	6 digital I/O ports
Kernel	Scan	Baud-rate sensitive	Read only

asynchronously by sending and receiving events. Within an active object, events are processed sequentially in a run-to-completion (RTC) fashion, while the scheduler encapsulates all the details of thread-safe event exchange and queuing.

As sensor programming is network-centric, we need to combine a reliable micro-kernel OS to build a framework for network-embedded OS. The main design goals are to design an adaptive MAC layer that needs to be interfaceable to many radio stacks and have dedicated tasks that sense periodic data from physical sensors in the background. Thus we design a sensor mote with two microcontroller units (MCUs). Dividing the functionality of sensing and communication can be accomplished with its own control. The requirements are easy programmability of the sensors and its control for any deployment. To have a microframework, we design a state machine object that allows the creation of active objects that can then send messages without blocking using a priority queue. This microframework can be extended to support microsecond accuracy using specific hardware abstraction layers (HALs) for available manufacturer sensor boards. There are several choices for programming the Texas Instruments (TI) MSP430. Here we try to abstract the TI architecture using a coresident small-footprint OS that allows the kernel booting into a mode f to have complete real-time control and at the same time be compatible with all the sensor interfaces that the MCU can control and configure. This allows the writing of sensing applications by creating tasks and timers and scheduling individual tasks on the part of the kernel.

The scheduler can be configured to have nonpreemptive/preemptive mode. *Non-preemptive scheduling* is when an intelligence surveillance reconnaissance (ISR) device is serviced and then the current task is run to completion. In the case of preemptive scheduling, the current task is rescheduled according to the priority after the ISR has completed, which might yield to another task already queued for processing.

12.1.4 MCU I Framework Specifications

MicroFrameWork (Table 12.2) is an event-driven C++ framework with its own multitasking kernel as shown in Fig. 12.1 aimed at the mote-class sensor networks. Its primary advantage over TinyOS [6] is that it allows modules to be easily ported into any processor that the sensor platform uses. Application-level software can be easily developed using a Unified Modeling Language (UML) diagram and the states as shown in Fig. 12.2, which represents emulation of an event model paradigm (see also Fig. 12.3).

```
Input: BatteryCapacity
Output: Residual Capacity
foreachEver do
    Sensor-ctor();
    SensorHardware-Init();
    /* initialize the board */
    MF-run();
    /* transfer control to MicroFrameWork*/
end
Return Data
```

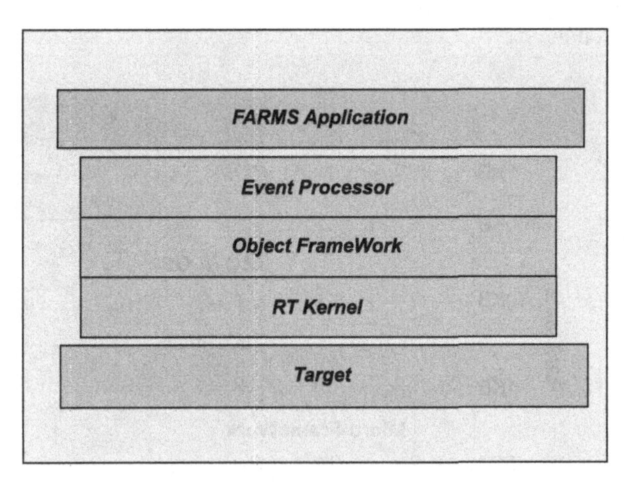

FIGURE 12.1 Stack architecture.

TABLE 12.2 Resource Requirements for Target

Software Platforms	ROM (kB)	RAM
DotNetMicroFrameWork	10	1 MB
Zotta OS	15	100 MB
MicroFrameWork	2	100 B
802.14.4 Network Stack	30–50	—

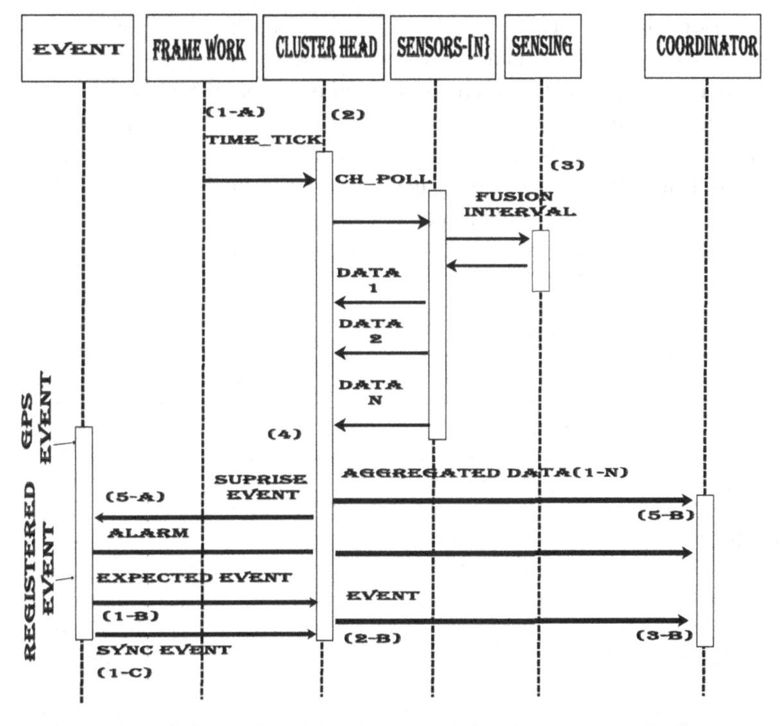

FIGURE 12.2 Active-object frame work for sensor communications.

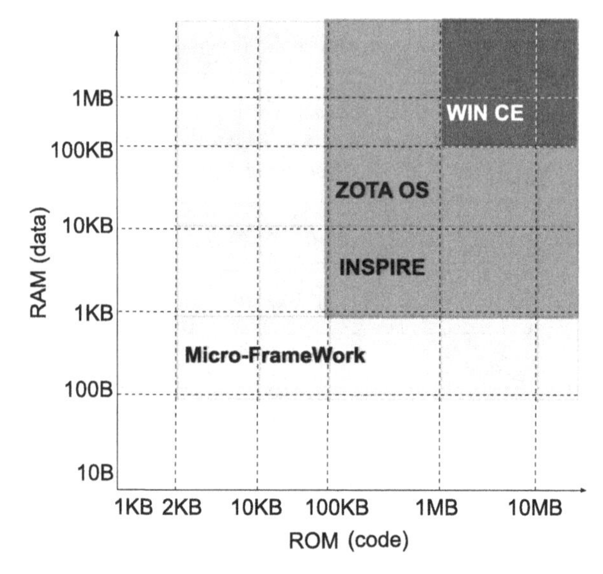

FIGURE 12.3 Total RAM and ROM size required for microframework, small RTOS, and other popular RTOSs and OSs.

12.1.5 MCU II Framework Specifications

In a sensor network stack (see Table 12.3), the network layer is solely responsible for route planning and maintenance. Most of the energy used by the network is due to its routing activity. Two methods are used in routing implementation: one at the network layer that is controlled by distributed algorithms to form clusters and uses efficient clusterhead selection and the other at the MAC layer that uses multihop routing to forward data at the lower layers by using best-effort quality of service (QoS).

As there are many different radios available, the need to make them work within a given framework is an integral part of the basic modular design. As the requirement

TABLE 12.3 STACK APIs

Protocol Stack	Fully Functional	Routing Stack	Monitoring Only
Layers	Stack	—	Stack
Application	`Form Network()`	`Do Routing()`	Join
Network	`Create Zones()`	`Elect ClusterHeads()`	`Measure API()`
Datalink	`CRF()`	`CRF()`	`CRF()`
Physical	IEEE Standard	IEEE Standard	IEEE Standard
OS scheduler	Create HostComm	Compress	`Create SenseTask()`
Microframework	—	—	`Create SenseTask 2`

is to develop a servicelike application that can control the MAC, we design all the core services needed to make it protocol-efficient. Any built-in features refer to "Optimized for power-aware OS." This is the case as the ultimate goal of the overall system in which any application is deployed is to have a longer lifetime.

```
void main( void )
    {

    //Coordinator Program
    //this initialization set
    //our SADDR to OxFFFF,
    //PANID to the default PANID

    //Hallnit, evblnit will have
    //to be called by the user
    hallnit();
    evblnit();

    apllnit();                    //init the stack
    conPrintConfig();
    ENABLE_GLOBAL_INTERRUPT();    //enable interrupts
    test_number =  0;

    //debug_level =  10;

    EVB_LED1_OFF();
    EVB_LED2_OFF();

    //get this for reference, will use the
    //LSB as srcEP for indirect message
    halGetProcessorIEEEAddress(&myLongAddress[0]);

#ifdef LRWPAN_COORDINATOR

    aplFormNetwork();

    //wait for finish
    while( apsBusy() )
        {
        apsFSM();
        }
    EVB_LED1_ON();
    conPrintROMString("Nwk␣formed\n");

#else
    do
        {
```

```
        aplJoinNetwork();

        while( apsBusy() )
            {
            apsFSM();
            } //wait for finish

        if( aplGetStatus() == LRWPAN_STATUS_SUCCESS )
            {
            EVB_LED1_ON();
            conPrintROMString("Network
␣␣␣␣␣␣␣␣␣␣␣␣␣␣␣␣␣␣␣␣␣␣␣␣␣␣␣␣␣␣Join␣succeeded!\n");
            conPrintROMString("My
␣␣␣␣␣␣␣␣␣␣␣␣␣␣␣␣␣␣␣␣␣␣␣␣␣␣␣␣␣␣␣ShortAddress␣is:␣");
            conPrintUINT16(aplGetMyShortAddress());
            conPCRLF();
            conPrintROMString("Parent␣LADDR:␣")
                    conPrintLADDR(aplGetParentLongAddress());
            conPrintROMString(",␣Parent␣SADDR:␣");
            conPrintUINT16(aplGetParentShortAddress());
            conPCRLF();
            break;
            }
        else
            {
            conPrintROMString("Network␣Join␣FAILED!
␣␣␣␣␣␣␣␣␣␣␣␣␣␣␣␣␣␣␣␣␣␣␣␣␣Waiting,then␣trying␣again\n");
            my_timer = halGetMACTimer();
            //wait for 2 seconds
            while( (halMACTimerNowDelta(my_timer))
                      < MSECS_TO_MACTICKS(2 * 1000) );
            }
        }  while( 1 );

#endif

#ifdef LRWPAN_RFD

    //announce ourselves to the coordinator
    //so that we can test indirect messaging
    //this is only necessary if there are routers
    // between us and the coordinator,
    //but since don't know the network topology,
    //do it always if RFD.

    do
         {
         //send to coordinator as it resolves bindings
         aplSendEndDeviceAnnounce(0);
```

```
        //wait for finish
        while( apsBusy() )
            {
            apsFSM( );
            }

        if( aplGetStatus() == LRWPAN_STATUS_SUCCESS )
            {
            conPrintROMString("End␣Device
␣␣␣␣␣␣␣␣␣␣␣␣␣␣␣␣␣␣␣␣␣␣␣␣␣␣␣␣␣␣Announc␣succeeded!\n");
            break;
            }
        else
            {
            conPrintROMString("End␣Device
␣␣␣␣␣␣␣␣␣␣␣␣␣␣␣␣␣␣␣␣␣␣␣␣␣␣␣␣␣␣Announce␣FAILED!
␣␣␣␣␣␣␣␣␣␣␣␣␣␣␣␣␣␣␣␣␣␣␣␣␣␣␣␣␣␣␣␣␣␣␣␣Waiting,␣then␣trying␣again\n");
            my_timer = halGetMACTimer();

            //wait for 2 seconds
            while( (halMACTimerNowDelta(my_timer))
                    < MSECS_TO_MACTICKS(2 * 1000) );
            }
        } while( 1 );
#endif

#if defined(LRWPAN_RFD) || defined(LRWPAN_COORDINATOR)

    //now send packets

    while( 1 )
        {
        packet_test();

        while( apsBusy() )
            {
            apsFSM();
            } //wait for finish
        }
#endif

#ifdef  LRWPAN_ROUTER

    //router does nothing, just routes

    conPrintROMString("Router,
␣␣␣␣␣␣␣␣␣␣␣␣␣␣␣␣␣␣doing␣its␣thing.!\n");
```

```
    while( 1 )
        {
        apsFSM();
        }
#endif

   }

␣␣␣␣␣␣␣␣␣␣␣␣␣␣␣␣␣␣␣␣␣␣␣␣␣␣␣␣␣␣␣␣␣␣␣␣␣␣␣␣␣␣␣␣␣␣␣␣␣␣␣␣␣␣␣␣␣␣
␣␣␣␣␣␣␣␣␣␣␣␣␣␣␣␣␣␣␣␣␣␣␣␣␣␣␣␣␣␣␣␣␣␣␣␣␣␣␣␣␣␣␣␣␣␣␣␣␣␣␣␣␣␣␣␣␣␣

void main( void )
    {

    //STACK API Program
    //this initializations
    // et our SADDR to OxFFFF,
    //PANID to the default PANID

    //Hallnit, evblnit will
    //have to be called by the user
    hallnit();
    evblnit();

    apllnit();//init the stack
    conPrintConfig();
    ENABLE_GLOBAL_INTERRUPT();
    //enable interrupts
    test_number = 0;

//debug_level = 10;

#ifdef LRWPAN_COORDINATOR

    aplFormNetwork();

    while( apsBusy() )
        {
        apsFSM();
        } //wait for finish

    conPrintROMString("Nwk␣formed,
␣␣␣␣␣␣␣␣␣␣␣␣␣␣␣␣␣␣␣␣␣waiting␣for␣join␣and␣reception\n");

    while( 1 )
        {
        apsFSM();
        }
```

```
#else

    do
        {
        aplJoinNetwork();

        while( apsBusy() )
            {
            apsFSM();
            }  //wait for finish

        if ( aplGetStatus()  ==
                    LRWPAN_STATUS_SUCCESS )
            {
            conPrintROMString("Network
␣␣␣␣␣␣␣␣␣␣␣␣␣␣␣␣␣␣␣␣␣␣␣␣Join␣succeeded!\n");
            conPrintROMString("My
␣␣␣␣␣␣␣␣␣␣␣␣␣␣␣␣␣␣␣␣␣␣␣␣␣ShortAddress␣is:␣");
            conPrintUINT16(aplGetMyShortAddress());
            conPCRLF();
            conPrintROMString("Parent␣LADDR:␣")
            conPrintLADDR(aplGetParentLongAddress());
            conPrintROMString(",␣Parent␣SADDR:␣");
            conPrintUINT16(aplGetParentShortAddress());
            conPCRLF();
            break;
            }
         else
            {
            conPrintROMString("Network␣Join
␣␣␣␣␣␣␣␣␣␣␣␣␣␣␣␣␣␣␣␣␣␣␣␣␣FAILED!␣Waiting,␣then␣trying␣again\n");
            my_timer = halGetMACTimer();

            //wait for 2 seconds
            while( (halMACTimerNowDelta(my_timer))
                   < MSECS_TO_MACTICKS(2 * 1000) );
            }
         }  while( 1 );

#ifdef LRWPAN_RFD

    //now send packets

    while( 1 )
        {
        packet_test();

        //wait for finish
        while( apsBusy() )
```

```
            {
            apsFSM();
            }
        }

#endif
#ifdef LRWPAN_ROUTER

    //router does nothing, just routes

    DEBUG_PRINTNEIGHBORS(DBG_INFO);
    conPrintROMString("Router,
␣␣␣␣␣␣␣␣␣␣␣␣␣␣␣␣␣␣␣␣␣␣doing␣its␣thing.!\n");

    while( 1 )
        {
        apsFSM();
        }

#endif

#endif

    }
```

12.1.6 WSN Sensing Applications

Sensing applications are defined with unique data sampling needs. To achieve this in a power-aware manner, a distributed WSN algorithm at the network layer is used to manage data routing from individual nodes to a central coordinator.

12.1.7 Data Routing

This category of routing does not have any application control at higher levels. As soon as the data are sampled, they are forwarded to the nearest forward node for delivery to the destination. The performance of such an application is based on more efficiently forwarding the data toward the destination.

```
void main(void){

  // Sensing device with reduced functions

      UINT32  my_timer;
      UINT8   failures, ping_cnt;

      //Hallnit, evblnit will
      //have to be called by the user
      hallnit();
      evblnit();
```

```
        aplInit(); //init the stack
        conPrintConfig();
        ENABLE_GLOBAL_INTERRUPT();
        //enable interrupts

        EVB_LED1_OFF();
        EVB_LED2_OFF();


        //debug, level = 10;

#ifdef LRWPAN_ROUTER
         routerState =
                ROUTER_STATE_JOIN_NETWORK;
         while(1)
         {
                apsFSM();
                switch( routerState)
                {
                 case ROUTER_STATE_JOIN_NETWORK:
                        aplJoinNetwork();
                        routerState = ROUTER_STATE_JOIN_WAIT;
                        break;
                 case ROUTER_STATE_JOIN_WAIT:
                        if(apsBusy()) break;
                        if(aplGetStatus() == LRWPAN_STATUS_SUCCESS)
                      {
                        conPrintROMString("Network...
␣␣␣␣␣␣␣␣␣␣␣␣␣␣␣␣␣␣␣␣␣␣␣␣Join␣succeeded!\n");
                        printJoin Info();
                        my_timer = halGetMACTimer();
                        routerState = ROUTER_STATE_NORMAL;
                        ping_cnt = 0;
                        failures = 0;
                        EVB_LED1_ON();
                      }
                      else
                      {
                        conPrintROMString("Network...
␣␣␣␣␣␣␣␣␣␣␣␣␣␣␣␣␣␣␣␣␣␣␣␣Join␣FAILED!␣Waiting,...
␣␣␣␣␣␣␣␣␣␣␣␣␣␣␣␣␣␣␣␣␣␣␣␣then␣trying␣again\n");
                        my_timer= halGetMACTimer();
                        //wait for 2 seconds
                        while((halMACTimerNowDelta(my_timer))...
                            < MSECS_TO_MACTICKS(2*1000));
                          routerState = ROUTER_STATE_JOIN_NETWORK;
                      }
                      break;

                 case ROUTER_STATE_NORMAL:
                        //check ping timeout
                        if(halMACTimerNowDelta(my_timer)
                              > MSECS_TO_MACTICKS(1000))
```

```
                {
                  //reset timer
                  my_timer= halGetMACTimer();
                  //long timeouts are done in
                  //segments of short intervals as
                  //the maximum timeout on the HAL
                  //MAC timer is platform dependent
                  ping_cnt++;

                  if(ping_cnt == PING_TIMEOUT)
                  {
                    ping_cnt = 0;
                    //send a ping
                    routerState= ROUTER_STATE_DO_PING;
                  }
                }
                break;
    case ROUTER_STATE_DO_PING:
            conPrintROMString("Sending_Ping !\n");
            //aplPingParent uses an APS ack in
            //order to ensure that this
            //node is still associated with the parent
            aplPingParent();
            routerState= ROUTER_STATE_WAIT_FOR_PING;
            break;
    case ROUTER_STATE_WAIT_FOR_PING:
            if(apsBusy()) break;
            if(aplGetStatus() == LRWPAN_STATUS_SUCCESS)
            {
                    //all is well
                    failures = 0;
                    my_timer = halGetMACTimer();
                    routerState= ROUTERSTATE_NORMAL;
            }else
            {
                    failures++;
                    conPrintROMString("Ping_failed!\n");
                    if(failures == MAX_PING_FAILURES)
                    {
                            failures = 0;
                            routerState =
                                 ROUTER_STATE_REJOIN_NETWORK;
                    }
                    break;
    case ROUTER_STATE_REJOIN_NETWORK:
            //A rejoin takes less
            //time than a join , and does
            //not erase the neighbor table.
            //A join erases
            //the neighbor table, forcing
            //all of the router's
            //childen to re-issue joins.
            EVB_LED1_OFF();
            //not connected to a network
```

```
                        aplRejoinNetwork();
                        routerState= ROUTER_STATE_REJOIN_WAIT;
                        break;
                case ROUTER_STATE_REJOIN_WAIT:
                        if(apsBusy()) break;
                        if(aplGetStatus() ==
                                LRWPAN_STATUS_SUCCESS)
                        {
                                my_timer = halGetMACTimer();
                                routerState = ROUTER_STATE_NORMAL;
                                ping_cnt = 0;
                                failures = 0;
                                EVB_LED1_ON();
                                conPrintROMString("Network
␣␣␣␣␣␣␣␣␣␣␣␣␣␣␣␣␣␣␣␣␣␣␣␣␣␣␣␣␣␣␣␣␣␣␣Rejoin␣succeeded!\n");
                                printJoin Info();
                                routerState = ROUTER_STATE_NORMAL;
                        }
                        else
                        {
                                failures++;
                                if(failures == MAX_REJOIN_FAILURES)
                                {
                                  conPrintROMString("Network␣Rejoin...
␣␣␣␣␣␣␣␣␣␣␣␣␣␣␣␣␣␣␣␣␣␣␣␣␣␣␣␣␣␣␣␣␣␣␣␣max␣tries␣exceeded.␣Trying␣a␣join.");
                                  routerState = ROUTER_STATE_JOIN_NETWORK ;
                                }
                                else
                                {
                                  conPrintROMString("Network␣ReJoin␣FAILED!...
␣␣␣␣␣␣␣␣␣␣␣␣␣␣␣␣␣␣␣␣␣␣␣␣␣␣␣␣␣␣␣␣␣␣␣␣Waiting,␣then␣trying␣again\n");
                                  my_timer= halGetMACTimer();
                                  //wait for 2 seconds
                                  while((halMACTimerNowDelta(my_timer))...
                                   < MSECS_TO_MACTICKS(2*1000));
                                  routerState = ROUTER_STATE_REJOIN_NETWORK;
                                  break;
                                }
                              }
                              break;
               default:
                       routerState = ROUTER_STATE_JOIN_NETWORK;
                       }
             }
        } //end while(1)
#else
        conPrintROMString("This␣application
␣␣␣␣␣␣␣␣␣␣␣␣␣␣␣␣is␣intended␣for␣a␣router!\n");
        conPrintROMString("Entering␣infinite
␣␣␣␣␣␣␣␣␣␣␣␣␣␣␣␣loop,␣doing␣nothing.\n");
        while(1);

#endif
}
```

12.1.8 Application with Sensing

This category of applications pools the sensors on a periodic basis, which allows average values to be accumulated over time. With this facility, the application can control the service usage, which further allows the service to adapt to predetermined data-sensing requests and also to redundant traffic.

```
void main( void )
    {
#ifndef LRWPAN_COORDINATOR

    UINT8 count;
    BOOL aps_ack;
    UINT8 failures;
    UINT32 my_timer;
#endif

    //HalInit, evbInit will
    //have to be called by the user

    halInit();
    evbInit();

    aplInit(); //init the stack
    conPrintConfig();

    ENABLE_GLOBAL_INTERRUPT();
    //enable interrupts

    EVB_LED1_OFF();
    EVB_LED2_OFF();
//debug, level = 10;

#ifdef LRWPAN_RFD

    rfdState = RFD_STATE JOIN_NETWORK;

    while( 1 )
        {
        apsFSM();

        switch( rfdState )
            {
            case RFD_STATE_JOIN_NETWORK:
                EVB_LED1_OFF();
                //not connected to a network
                aplJoinNetwork();
                rfdState = RFD_STATE_JOIN_WAIT;
                break;

            case RFD_STATE_JOIN_WAIT:
                if( apsBusy() )
                    break;
```

```
        if( aplGetStatus() == LRWPAN_STATUS_SUCCESS )
            {
            conPrintROMString("Network
␣␣␣␣␣␣␣␣␣␣␣␣␣␣␣␣␣␣␣␣␣␣␣␣␣␣␣␣␣␣Join␣succeeded!\n");
            printJoinInfo();
            rfdState = RFD_STATE_NORMAL;
            ping_cnt = 0;
            count = 0;
            aps_ack = FALSE;
            EVB_LED1_ON();
            }
        else
            {
            conPrintROMString("Network
␣␣␣␣␣␣␣␣␣␣␣␣␣␣␣␣␣␣␣␣␣␣␣␣␣␣␣␣␣␣␣␣␣␣␣Join␣FAILED!␣");
            conPrintROMString("Error:␣");
            conPrintUINT8(aplGetStatus());
            conPrintROMString(",␣Waiting ,
␣␣␣␣␣␣␣␣␣␣␣␣␣␣␣␣␣␣␣␣␣␣␣␣␣␣␣␣␣␣then␣trying␣again\n");
            my_timer = halGetMACTimer();

            //wait for 2 seconds
            while( (halMACTimerNowDelta(my_timer))
                   < MSECS_TO_MACTICKS(2 * 1000) );
            rfdState = RFD_STATE_JOIN_NETWORK;
            }
        break;

    case RFD_STATE_NORMAL:
        //send to some target in the tree.
        dstADDR.saddr = PING_DST_SADDR;
        //send a message, then sleep
        //increment ping counter
        ping_cnt++;
        count++;

        //every so often, send an APS ack
        // request to ensure that we are
        //still actually associated with
        // our parent and that the packet
        //actually reached the coordinator.
        // APS acks require
        //more waiting time and overhead,
        // so only use them when necessary.
        //A MAC ack is always requested for
        // a data packet, but this only
        //ensures that the packet was
        // received by the radio of our parent
        //(ie, we have the correct short
        // address/panid of the parent).
        //if the parent has dropped us from its
        // neighbor table for some
        //reason, then the packet is rejected
        // at at the nwk level.
```

```
                //Also, if we are going through a
                //router(s) to the coordinator,
                //then the MAC ack is only good for
                //the first hop to our parent.
                If( count == 4 )
                    {
                    conPrintROMString("Requesting
                                                   APS_ack\n");
                    aps_ack = TRUE;
                    count = 0;
                    }
                 else
                    {
                    aps_ack = FALSE;
                    }
                 payload[0] = (BYTE)ping_cnt;
                 payload[1] = (BYTE)(ping_cnt >> 8);

                 //This uses an APS ACK so that
                 //know if the message
                 //was delivered. If it fails ,
                 //then we assume that
                 //we have lost connection, and
                 //we issue a join
                 aplSendMSG(APS_DSTMODE_SHORT,
                     &dstADDR, 2, //dst EP
                 0,//cluster is ignored for direct message
                 1,//src EP
                 &payload[0], 2, //msg length
                 apsGenTSN(), aps_ack);
                 rfdState = RFD_STATE_WAIT_FOR_ACK;
                 break;

            case RFD_STATE_WAIT_FOR_ACK:
                 if( apsBusy() )
                     break;

                 if( (aplGetStatus() ==
                     LRWPAN_STATUS_SUCCESS) || ! aps_ack )
                     {
                     //all is well
                     rfdState = RFD_STATE_SLEEP;
                     }
                 else
                     {
                     //only try a rejoin if the aps_ack failed.
                     //if mac_ack failed, we will keep
                     //trying until the aps_ack fails.
                     //we assume that we have been disconnected.
                     //Try rejoining first , then a join.
                     failures = 0;
                     rfdState = RFD_STATE_REJOIN_NETWORK;
                     }
                 break;
```

```
            case RFD_STATE_SLEEP:

                 conPrintROMString("Going␣to␣sleep...\n");
                 //This does a disable global interrupt!
                 aplShutdown();
                 halWaitMs(IO);
                 //Going to sleep is
                 // platform/application dependent.
                 //the halSleep function is
                 // only intended for example purposes
                 // the msecs argument in halSleep
                 // may be ignored by the HAL layer
                 //as how sleep is implemented is
                 // target dependent.
                 halSleep(4000);
                 conPrintROMString("Woke␣up!\n");
                 aplWarmstart();
                 ENABLE_GLOBAL_INTERRUPT();
                 rfdState = RFD_STATE_NORMAL;
                 break;

#ifndef LRWPAN_COORDINATOR

            case RFD_STATE_REJOIN_NETWORK:
                 conPrintROMString("Trying␣to
␣␣␣␣␣␣␣␣␣␣␣␣␣␣␣␣␣␣␣␣␣␣␣␣␣␣␣rejoin␣network!\n");
                 aplRejoinNetwork();
                 rfdState = RFD_STATE_REJOIN_WAIT;
                 break;

            case RFD_STATE_REJOIN_WAIT:
                 // if stack is busy, continue
                 if( apsBusy() )
                      break;

                 if( aplGetStatus() ==
                           LRWPAN_STATUS_SUCCESS )
                      {
                      failures = 0;
                      EVB_LED1_ON();
                      conPrintROMString("Network
␣␣␣␣␣␣␣␣␣␣␣␣␣␣␣␣␣␣␣␣␣␣␣␣␣␣␣␣␣␣Rejoin␣succeeded!\n"};
                      printJoinInfo();
                      rfdState = RFD_STATE_NORMAL;
                      }
                 else
                {
                 failures++;

                 if( failures ==
                       MAX_REJOIN_FAILURES )
                 {
                 //this starts everything over
```

```
                 conPrintROMString("Max Rejoins
                                         failed, trying to join \n"};
                 rfdState = RFD_STATE JOIN_NETWORK;
                 }
                 else
                 {
                   //else, wait to try again
                   conPrintROMString("Network
                 Rejoin FAILED!
                                              Waiting, then trying again\n");
                   //most apps probably
                   // do not need to
                   //wait before retrying
                   //rejoin, this is included
                   // just for visibility
                   //purposes in reading the output
                   my_timer = halGetMACTimer();

                   //wait for 2 seconds
                   while( (halMACTimerNowDelta(my_timer))
                    < MSECS_TO_MACTICKS(2 * 1000) );
                       rfdState = RFD_STATE_REJOIN_NETWORK;
                  }
                 }
                 break;

#endif

             default:
                 rfdState = RFD_STATE_JOIN_NETWORK;
             }
         }
#else

    aplFormNetwork();

    //wait for finish
    while( apsBusy() )
        {
        apsFSM();
        }
    conPrintROMString("Network formed,
                                                       waiting for RX\n");
    EVB_LED1_NO;

    //coordinator or router just runs the stack
    while( 1 )
        {
        apsFSM();
        }

#endif

    }
```

12.1.9 Ultralow Duty Cycling Using FARMS

When energy resources are abundant, due to the availability of renewable energy resources, the running harvesting application is sensitive to the recharge rate. Since this allows the transmission of data to a forwarding node, the communication need not be synchronized. The duty cycle still needs to periodically poll the channel to check for any activity. For reliable delivery, a sufficient number of receivers should be active to ensure that no forwarded packets are dropped. This feature needs to be scheduled since receiving and idling can drain all the nodes in a given area if they are not periodically scheduled.

12.1.10 Real-Time System Components

To build a predictable system, *all* of its wireless sensor network components, hardware and software, plus a good design, contribute to this long-term predictability. Having both good sensor hardware and a good networkable RTOS is a minimal but insufficient requirement for building a correct reactive sensor system. An improperly designed testbed system with excellent hardware and software building blocks may still lead to disaster. This section deals with sensor testbed microframework. In general, a good microframework can be defined as one that has a bounded (predictable) behavior under all system load scenarios [1]. The current architecture uses a more fiexible design with two microcontrollers as discussed before, one dedicated to transmission and the other to data sensing. As a scheduler is more dependent on sensing, a real-time framework is used to create specific sensing tasks that allow for the isolation of error during sensing and transmission.

12.1.11 Complexity

Wireless sensor systems are dedicated and unattended for most of their lifetimes, so the complexity of the design needs to address the time (scheduling, polling, synchronization) local, global, and communication overhead to generate the datastream (real-time sensor data). We use specific MAC-level protocols such as Berkeley MAC (B-MAC) [4] used by TOSSIM, carrier sense multiple access (CSMA), and the general IEEE 802.11 wireless predecessor without RTS/CTS. These are implemented using the GlomoSIM routing layer.

12.1.12 Event-Driven System

Wireless sensor systems are dedicated and event-driven [1,2]. We use a common platform that allows modules to be programmed with events. A simple sensor cluster is implemented in Fig. 12.2 that has five states for performing a useful task.

Poll is generated by the microframework, which allows tracking of the current hardware clock. Once the poll is active, the sensor object changes states. If it is idle, then it will accept the poll and initialize itself with a predetermined timeout. The timeout event is handled, and the measured value from the sensor is read accurately

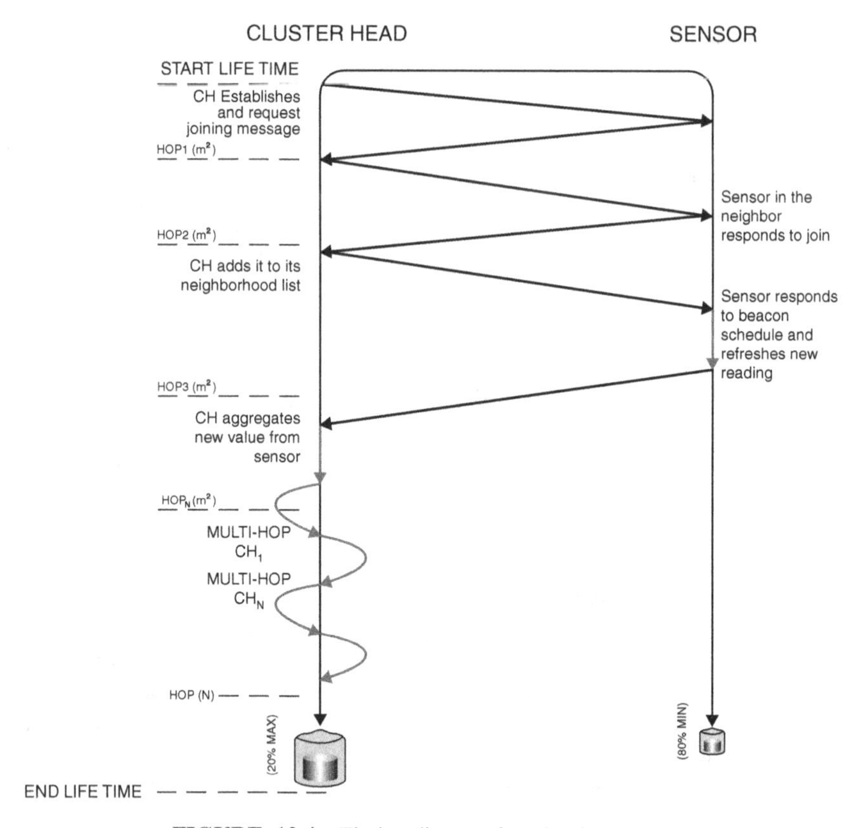

FIGURE 12.4 Timing diagram for a hardware mote.

and processed. If there is a threshold set for this particular process, then an alarm will be enabled according to the current read value. The on-idle event can support the low-power features of the target hardware, enabling further energy savings. Features of poll and on-idle events are presented in Table 12.4. Both poll and on-idle events are characterized.

12.2 SOFTWARE MICROFRAMEWORK REQUIREMENTS

The heart of the design is the event-driven microframework, which allows the building of robust, reliable, and resilient sensor motes. Here the framework uses well-defined

TABLE 12.4 Sensor-Processing Events

Poll	Initialize	Process
On-idle	Timeout	Event/alarm

object-oriented techniques, which are easy to maintain, and also enables modeling state machines. These machines are predictable and have deterministic state transition during their entire lifetime execution.

12.2.1 State Machine

If you have seen existing design and codes that are designed for various parts of the network stack, you may have observed that they are riddled with a disproportionate number of convoluted conditional execution branches (deeply nested if–else or switch-case statement in C/C++). This highly conditional code is a testament to the basic characteristics of reactive systems. If a redesign could eliminate a fraction of these conditional branches, the code would be much easier to understand and test, and the sheer number of convoluted execution paths through the code would drop radically, perhaps by orders of magnitude. Techniques based on state machines are capable of achieving exactly this dramatic reduction of the different paths through the code and simplification of the condition tested at each branching point [3,7]. The state machines described in the cluster timing UML specification represent the current state of the art in the long evolution of these techniques.

12.2.2 Sensor State Machine/UML Diagram (Algorithm)

A system exhibits state behavior when it operates differently during different periods, and its behavior can be partitioned into finite, nonoverlapping chunks called *states*. For example, basic mathematical functions, such as sin(x), return the same result for a given input x regardless of the history of previous inputs x_i. A common, straightforward way of modeling state behavior is through a Finite State Machine (FSM). Using FSMs is an efficient way to specify constraints of the overall behavior of a system. Being in a state where the system responds to only a subset of all allowed inputs produces only a subset of possible responses and changes state directly to only a subset of all possible states. Figure 12.2 shows the active objects that constitute a networked sensor embedded system. Active objects in a micro framework are encapsulated tasks (each embedding a state machine and an event queue) that communicate with one another asynchronously by sending and receiving events. For active objects within the framework, events are processed sequentially in a run-to-completion (RTC) fashion, while the scheduler encapsulates all the details of thread-safe event exchange and queuing.

REFERENCES

1. C. G. Cassandras and S. Lafortune, *Introduction to Discrete Event Systems*, Kluwer Academic, Jan. 1999.
2. G. S. Fishman, *Principles of Discrete Event Simulation,* Wiley, 1978.
3. K. Ilgun, R. A. Kemmerer, and P. A. Porras, State transition analysis: A rule-based intrusion detection approach, *IEEE Trans. Software Eng.* **21**(3):151–180 (March 1995).

4. V. Iyer, R. M. Garimella, and M. B. Srinivas, Min loading max reusability fusion classifiers for sensor data model, *Proc. 2nd Int. Conf. Sensor Technologies and Applications, SENSORCOMM '08,* Aug. 25–31, 2008, pp. 480–485.
5. V. Iyer, G. Rama Murthy, and M. B. Srinivas, Training data compression algorithms and reliability in large wireless sensor networks, *Proc. IEEE Int. Conf. Sensor Networks, Ubiquitous, and Trustworthy Computing,* 2008, pp. 480–485.
6. J. Polastre, J. Hill, and D. Culler, Versatile low power media access for wireless sensor networks, *Proc. 2nd Int. Conf. Embedded Networked Sensor Systems, SenSys '04,* ACM, New York, 2004, pp. 95–107.
7. M. G. Schultz, E. Eskin, et al., Data mining methods for detection of new malicious executables, *Proc. IEEE Symp. Security and Privacy, May* 2001.
8. W. Ye, F. Silva, and J. Heidemann, Ultra-low duty cycle MAC with scheduled channel polling, *Proc. 4th Int. Conf. Embedded Networked Sensor Systems, SenSys '06,* ACM, New York, 2006, pp. 321–334.

13 Performance Analysis of Power-Aware Algorithms*

Efficiency and quality are of equal importance!! Both come from experience, not from study. Study as you go, don't assume that you're ready for the real world because you studied first.

—Jon Davis

13.1 INTRODUCTION

In the previous chapter most of the application development framework was been discussed for specific deployment, such as the IEEE 802.15.4 ZigBee Alliance. These frameworks make application development and deployment very transparent to developers that hide some of the unique performance characterstics of sensor networks. In this chapter, we will use some of the parameters that are needed for scalable applications such as node counts in the neighborhood of 100 and power dissipation and loading of traffic at each node for different algorithms. This treats the large network as a testbed with standard metrics used by life time and standard AA cells. The testbed simulator allows us to compare the performance characteristics of different hardware, MAC protocols, and their energy requirements.

13.1.1 Performance Metrics

We can define *lifetime of sensor networks* as

$$\text{Node lifetime} = \frac{C_{\text{batt}} \times V \times 60 \times 60}{E_{\text{Tx}+\text{Rx}+\text{idle}+\text{Lis}}} \times 100\% \tag{13.1}$$

where C = power consumption, E = energy, V = voltage, Rx/Tx = receving/transmitting, and $\text{Lis}_{\text{passive}}$.

Idle power consumption is defined as the time elapsed between request and receipt of information over the wireless channel. This is increasing at an alarming rate, due

*Portions of this chapter were contributed from various sources by V. Iyer [1–4].

Fundamentals of Sensor Network Programming: Applications and Technology, By S. S. Iyengar, N. Parameshwaran, V. V. Phoha, N. Balakrishnan, and C. D. Okoye

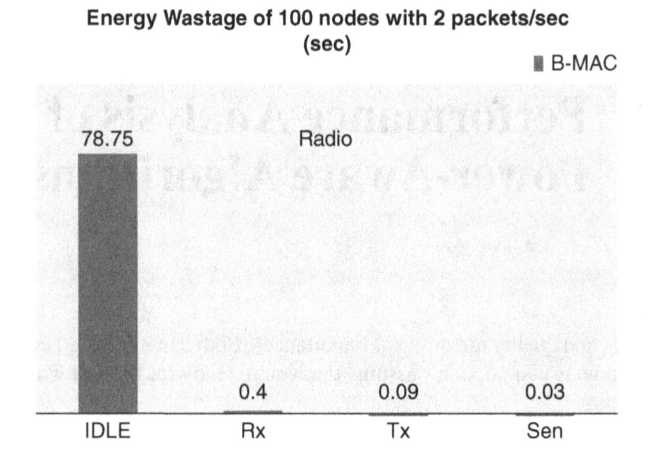

FIGURE 13.1 Energy wastage due to radio using Eq. (13.1).

to exponential growth of idling. Figure 13.1 illustrates the simple WSN application that aggregates data with 100 nodes that must run the radio of all the nodes all the time to establish the network path. Along that path are numerous router nodes and clusterheads. Each of these nodes helps manage wireless traffic to piggyback the data to the destination, creating a multihop page between the data source and the destination node. When a new data update is requested, all of its neighboring nodes participate to enable the data to travel separately through this maze of the networked nodes. Each "hop" along the way is an opportunity to introduce power savings and thus add to optimizing the overall power consumption of the sensor network.

While manyfactors contribute to energy wastage, unscheduled overhearing and channel polling represent the largest and fastest-growing portion of total energy wastage. In fact, studies [1] have shown that 80% of the time nodes in a dense sensor network are affected by local ambient conditions such as collision, overhearing, and control packets.

The example shown in Fig. 13.2 illustrates the harmful effect of even a small degree of energy wastage during idling under ideal conditions. When lifetime decreases, such as during transmittance or when packet loss rates increase, performance can significantly deteriorate. In fact, studies have shown that it is more efficient for a routing algorithm to use a minimal number of nodes such as cluster heads or multihop leader nodes to schedule its data delivery.

One way to avoid the negative effects of "(idle wastage) × (number of nodes)" on life time performance is to have receiver-centric duty cycling for the majority of the nodes, thereby minimizing the number of active required nodes in the connected network. That is precisely how the INSPIRE runtime framework process works, as shown in the Eq. (13.2), with the duty cycle as shown in Fig. 13.2.

$$\text{Node lifetime} = \frac{C_{\text{batt}} \times V \times 60 \times 60}{E_{\text{Tx}+\text{Rx}+\text{idle}+\text{Lis}}} \times \text{duty cycle} \tag{13.2}$$

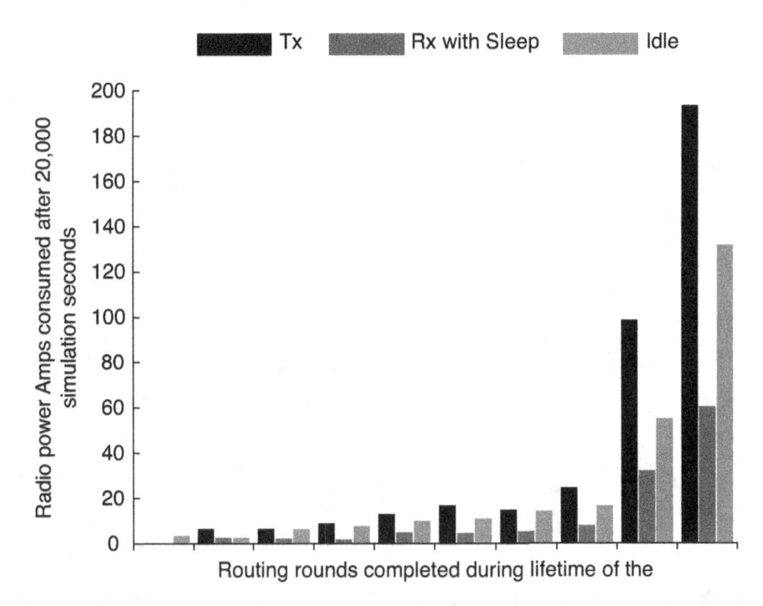

FIGURE 13.2 Unbalanced energy power resource allocations due to radio using Eq. (13.2).

The INSPIRE process utilizes several innovative technologies to extend the lifetime of WSN applications. One advantage of these techniques that specifically addresses unbalanced resource allocation for a proactive application is that it is very sensitive to Tx energy cost. As noted in Fig. 13.1, the Tx peaks during the application lifetime. This runtime environment (FARMS) uses a renewable energy model that attempts to average the resource allocation over the entire lifetime for ultra-low-duty-cycle applications, so we rewrite Eq. (13.2) to use ultralow duty cycling as shown in Eq. (13.3) and add a rechargeable cycle to renew ambient energy as given in Eq. (13.4), thus making new applications significantly longer to enable continuous collection of data:

$$\text{Node lifetime} = \frac{C_{\text{batt}} \times V \times 60 \times 60}{E_{\text{Tx}+\text{Rx}+\text{idle}+\text{Lis}}} \times 0.01\% \qquad (13.3)$$

$$\begin{array}{l}\text{Node lifetime for renewable}\\ \text{energy routing nodes}\end{array} = \frac{\text{recharge rate (mAh)}}{\text{Fixed Tx cost (mAh)}} \qquad (13.4)$$

These services share the resources of a node found among many fusionable ambient renewable MACS(FARMS) [1] deployed to route the data with the available software, hardware, and compatible radios as shown in Fig. 13.3. In particular, FARMS provides an innovative way to describe the reactive properties of the sensor network when sensing in a real-time environment. Evolution of the INSPIRE framework allows us to port distributed algorithms with a transparent power-aware service meant for sensing applications.

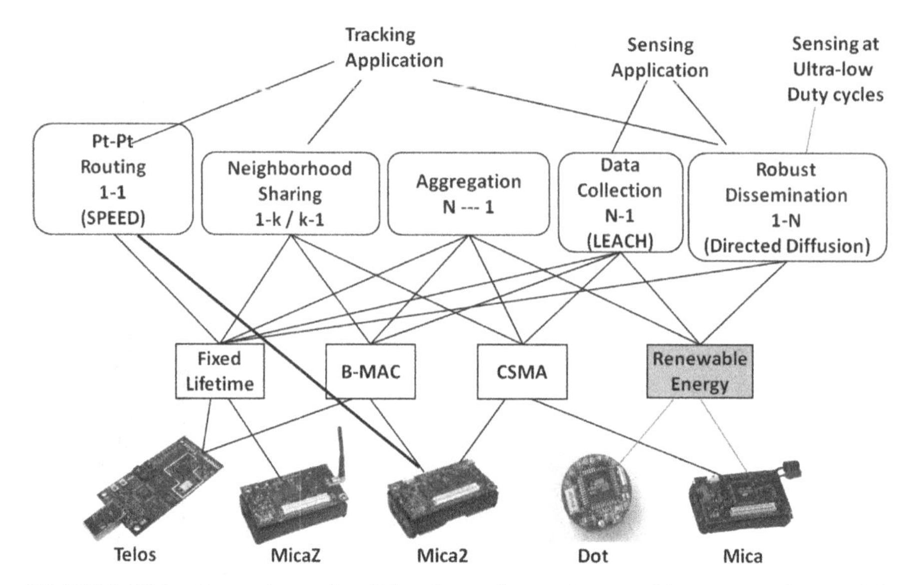

FIGURE 13.3 Comparison of traditional, reactive, and renewable programming models using compatible wireless radios.

13.2 SERVICE ARCHITECTURE

Stack architecture, service architecture, stack and service implementations, and related details are shown in Figs. 13.4–13.11.

As in the previous work, most of the simulation [3] has been done using a programming stack model, and here, in this architecture, as compared in Fig. 13.3, a multihop

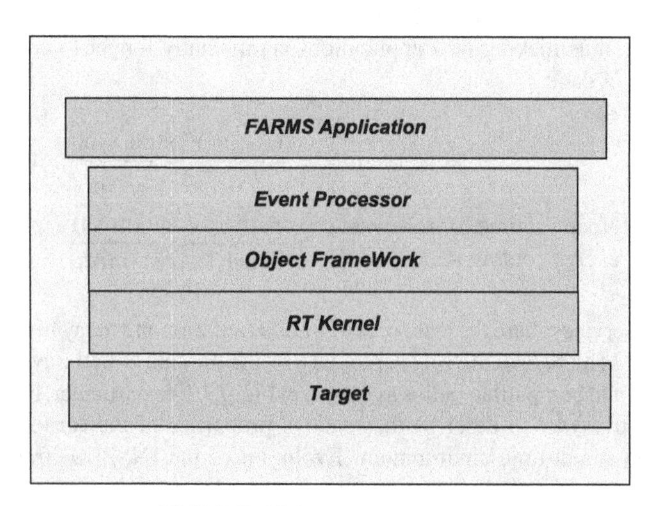

FIGURE 13.4 Stack architecture.

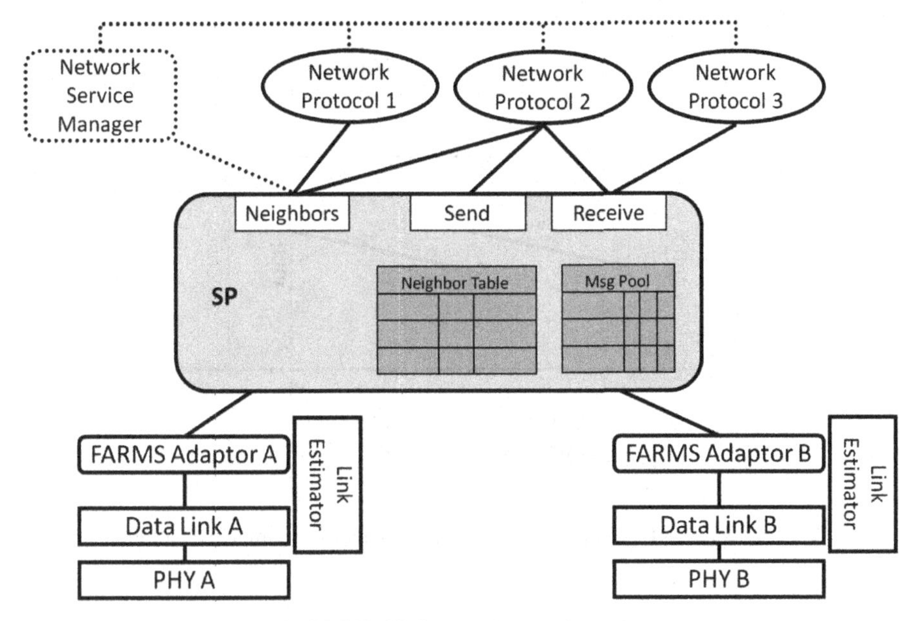

FIGURE 13.5 Service architecture.

data-forwarding service is emphasized. Also, from the traditional distributed model, which calculates lifetime using fixed resources, we extend the constrained resources by providing a dynamic renewable energy resources model. We essentially refer to this process as "adapting to the ground truth of an ambient wireless phenomenon." In addition, the stack model supports the Rx, idle, recharge cycle, and energy wastage

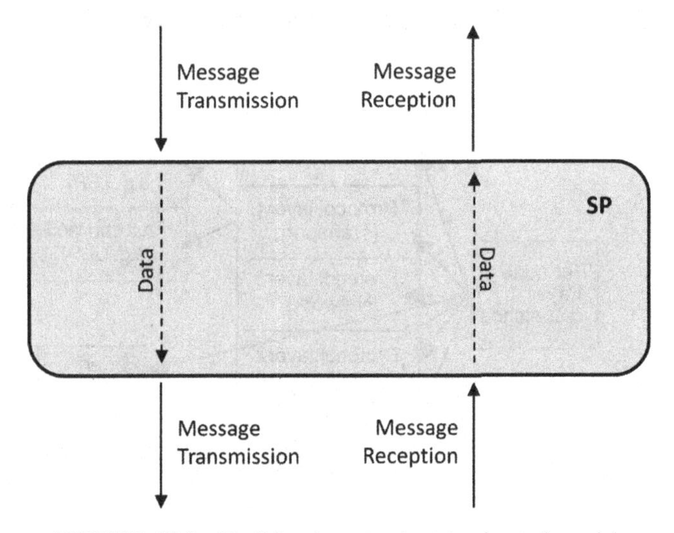

FIGURE 13.6 Traditional opaque layering in stack model.

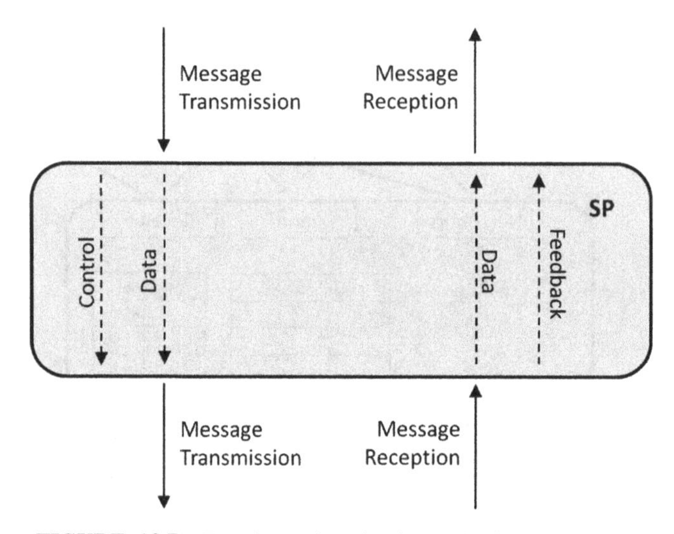

FIGURE 13.7 Translucent layering in service implementation.

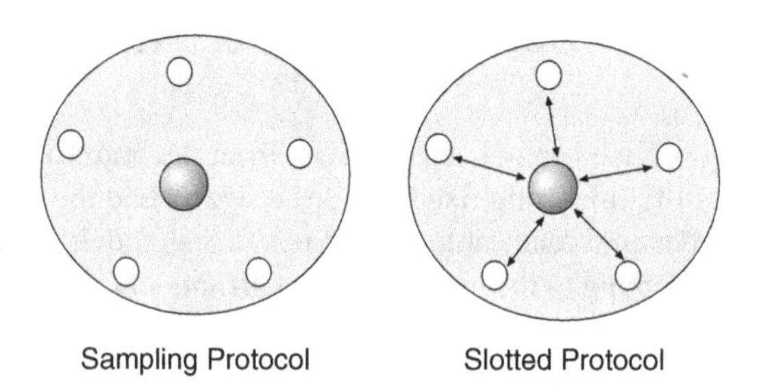

FIGURE 13.8 Clustering data aggregation.

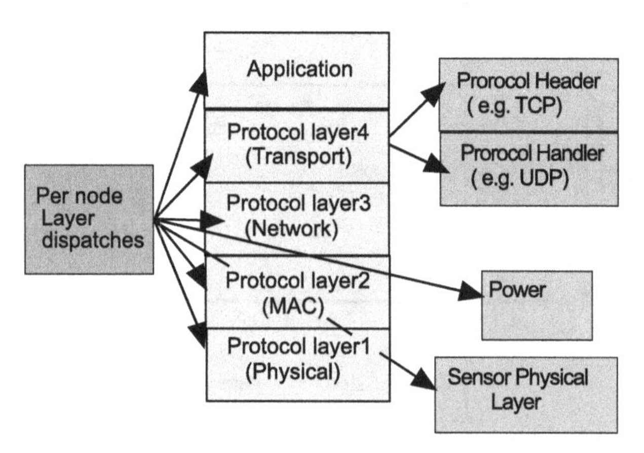

FIGURE 13.9 GlomoSIM architecture.

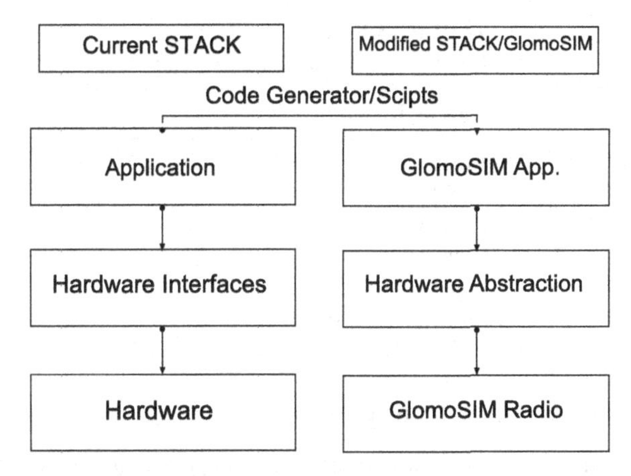

FIGURE 13.10 Application code development cycle.

due to the node deployment density ratio. The key elements of the service protocol design are listed in Tables 13.1 and 13.2.

13.2.1 Reliability

The traditional concept of link abstraction does not hold true for WSN as it not only communicates wirelessly but also performs data sensing, which is prone to outside (spurious) noise and channel interference. Due to this design constraint, the link quality is defined to detect good data from occurring noise levels. This helps application developers rely on the data being measured and validates the significance of the data to the deployed application when running unattended.

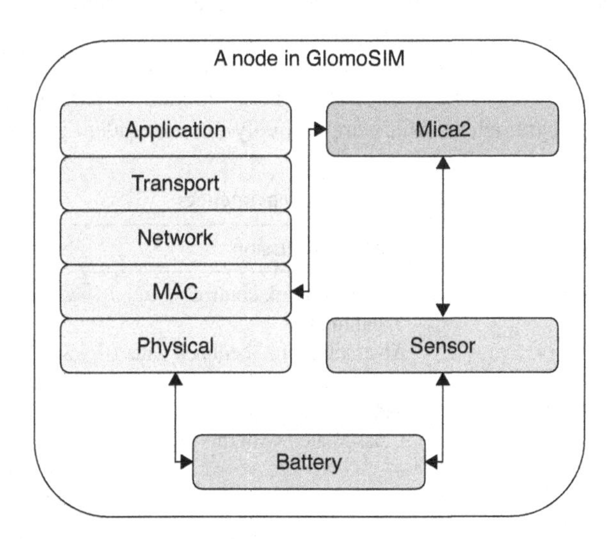

FIGURE 13.11 Simulator framework.

TABLE 13.1 Data-Gathering Reliability for Large Sensor Network

Sensing Reliability
Acknowledgments/ARQ
RTS/CTS
Priority
Congestion control
Fragmentation
Link quality estimation

13.2.2 Communication Transparency

The basic internode–intercluster communication paradigm needs to be very efficient in terms of network latency and power savings, and be able to adapt to traffic changes. These parameters, which are at the runtime framework level and independent of the application, are listed in Table 13.2. A simple implementation uses leveled priority or standard queue to manage message pools. As this subtopic is sensitive to node placement and harsh deployment conditions, we assume a clusterlike topology of physical nodes that are balanced using a neighborhood table to schedule the message to forward clusters by minimizing communication overheads. Refer to Figs. 13.12b and 13.12c for multihop and single-hop performance during application lifetime. The service framework power savings optimization with deployment of node density in lifetime is shown in Fig. 13.12d.

A service-driven architecture approach defines the minimal set of abstraction primitives that are based on distributed WSN runtime conditions that make it transparent to the programmer. Given the increasing importance of optimal quality of service in enhancing the lifetime of sensor networks, the converse problem of energy equivalence in routing is equally significant and not yet addressed in the literature in a quantifiable manner. We address both of these issues (QoS) by modeling a fixed-lifetime resource model (related work) and a renewable energy model. In particular, we extend the common QoS parameters, which are uniquely interdependent in both the models, to

TABLE 13.2 Wireless Communication Dependences

Data Reception	Data Transmission	Neighbor Management
Message arrives from link	Abstracted link control parameters	Cooperatively managed
Service dispatches	Abstracted link feedback data	Service mediates interaction using table
Network protocols establish	References to packets associated with this message	No policy on admission/ eviction by HP
Naming/addressing	—	Link Power Scheduling information
Filtering	—	—

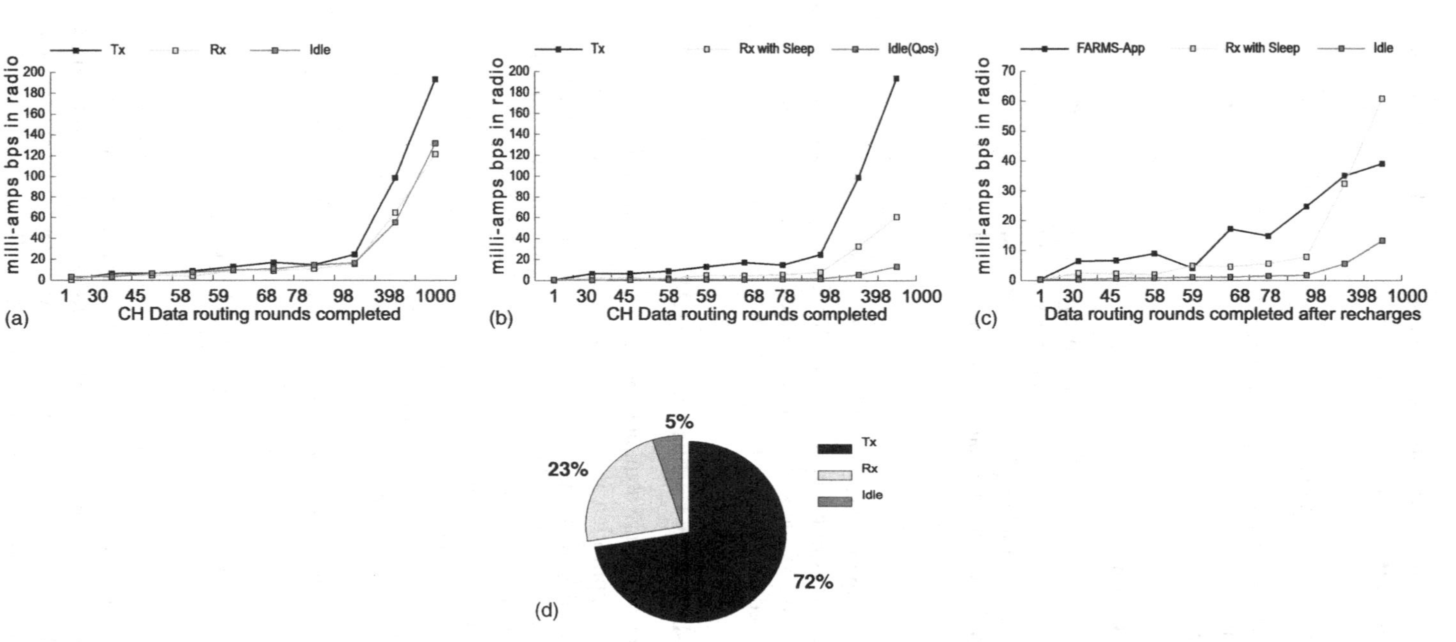

FIGURE 13.12 MAC performance comparison for normal and renewable lifetimes: (a) MAC performance with data forwarding; (b) MAC performance with 20% nodes as clusterheads with application control; (c) FARMS application showing lower transtion of data in MAC Layer; (d) sensor task overhead breakdown for network and MAC layers combined.

study the effects of ultra-low-duty cycling applications. The QoS application of the service can use the predefined data delivery.

13.2.3 Minimal Application

How to implement software in sensor networks is essential to understanding sensor networks. A network architecture and protocols are essential foundations for building software applications.

13.2.4 Data Routing

This category of routing does not have any application control at the higher level. As soon as the data are sampled, they are forwarded to the nearest forward node to be delivered to the destination. The performance of such an application is based on more efficiently forwarding the data toward their destination.

13.2.5 Application with Sensing

This category of applications pools the sensors on a periodic basis, which then allows average values to be accumulated over time. As a result of this feature, the application can control the service usage, which further allows the service to adapt to a predetermined data-sensing request and adapt to redundant traffic.

13.2.6 Ultra-Low-Duty Cycling Using FARMS

When energy resources are abundant owing to renewable energy resources, the running harvesting application is sensitive to the recharge rate. As this allows it to transmit data to a forwarding node, the communication need not be coordinated; however, this feature enables the application to ensure that the channel is available. For reliable delivery there should be a sufficient number of receivers active so that no forwarded packets are dropped. This needs to be scheduled as receiving and idling can drain all the nodes in a given area if they are not actively scheduled.

This development framework allows us to implement a very optimized version of routing techniques with specific data sensing needs and allows the framework to be transparent to the developer. For details on the performance comparison of the power consumption for each application category, see Figs. 13.12–13.14.

13.3 APPROACHES TO WSN PROGRAMMABILITY

13.3.1 GlomoSIM

GlomoSIM, a commercially available networks imulator, combines the simulation kernel, which can be run on a dual-core PC. GlomoSIM scales well with the number of nodes. We have tested between 100 and 500 nodes set up randomly in a 420 × 420-m grid. Its strength lies in this scalability as well as its exceptional radio propagation model. GlomoSIM was chosen as the base simulator because its framework

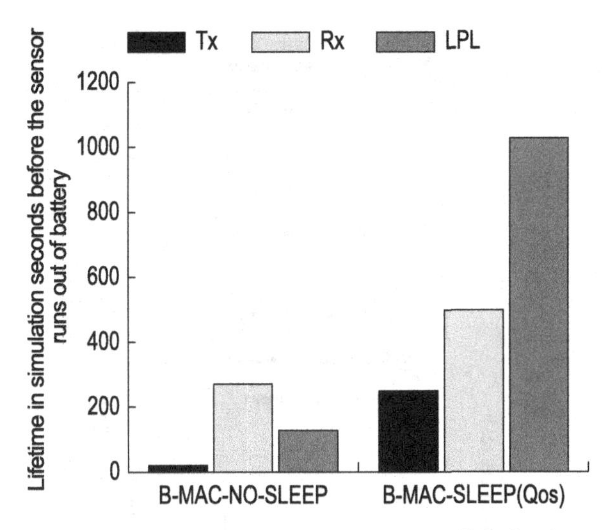

FIGURE 13.13 Scheduling idle with sleep and low-power listening increases lifetime by 2 times.

is superior to that of NS-2 and other network simulators. GlomoSIM allows new models to be added to any layer with ease as shown in Fig. 13.9. The basic means of communication for multihop or internode communication is their message-passing APIs. An application generates a message to be sent out of the radio using the same API as it uses to set a timer for itself. This independence of the message-passing API from the simulation leads to simpler programming.

The strength of Glomo SIM is its scalability and exceptional propagation models. This is important because of two factors:

1. *Scalability*. When simulating sensor networks, it is not uncommon to simulate tens of thousands of nodes. While a traditional network might be simulated with only 100s of nodes, a sensor network is inherently larger in size. A simulator that is designed for use as a sensor network simulator must be able to simulate the larger number of nodes.
2. *Propagation models*. When the user simulates a large sensor network, the topology used might be that of an actual deployment. It is, therefore, very important that the wireless channel be modeled accurately. The simulation can then be used to detect connectivity and other such problems.

13.4 SIMULATION CAPABILITIES

13.4.1 Unmodified Code Simulation

One primary focus of this implementation is to allow FARMS applications to be simulated inside GlomoSIM. Forcing the programmer to make GlomoSIM-specific [3] modifications in the code decreases from the ease of use of the simulator as shown

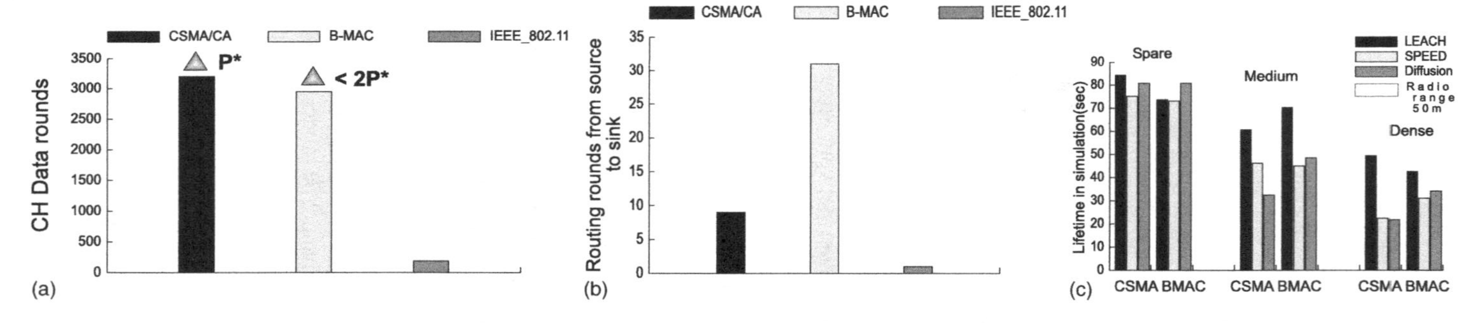

FIGURE 13.14 MAC performance comparison for data aggregation and multihop algorithms using SPEED (systems planning, engineering, and evaluation device), directed diffusion, and LEACH (low-energy adaptive clustering hierarchy): (a) local data aggregation protocol performance at clusterhead; (b) multihop protocol performance from source to sink; (c) MAC performance for sparse, medium, and high-density deployment with constant radio range.

TABLE 13.3 Time Scheduled by MAC During Useful Network Lifetime

Power-Aware MAC	Tx	Rx	Idle	QoS
CSMA	1%	80%	19%	Implemented by nonsensor network simulators
B-MAC	2%	65%	33%	TinyOS (without real-time clock)
Application control single-hop MAC	6%	45%	49%	Our study and first adaptation of power-aware QoS
Time scheduling FARMS multihop MAC	72%	5%	23%	Extended version tunable close to design constraint at idle ($\leq 1\ \mu$Ah)

in Fig. 13.11. Therefore, all necessary code porting is performed via a parser/code generator and script files. That the programming language is C allows for a much simpler code generator.

13.4.2 Sensor Stack

The sensor stack [4] can be simulated on the real node as shown in Fig. 13.12, thus allowing for real data to be generated. This can be used for various purposes, including verification of a sensor model. One caveat is that an application that requires strict timing for its data would not be accurate with this mode of simulation. This mode is designed for applications that process sensor data with lax timing. One such application is data gathering. Most applications generate data until their buffer is filled and only then do they transmit the data. This mode can also work if the application requires samples at a fairly low rate, such as 10 samples per second (sps).

13.4.3 Channel Emulation

The MAC and radio layer use the modified version of the GlomoSIM radio. This is very useful as it combines the best of simulations with the best of the real-world test bed.

13.5 BENCHMARKING

Table 13.3 and Figures 13.13 and 13.14 present a detailed result of programmable simulations of cross-layer structures of sensor networks; more specifically, the results highlight the time scheduled by MAC during useful network lifetime. For more details on implementation, refer to Ref. 1.

For futher details on benchmarking, refer to Refs. 1 and 2.

13.6 CONCLUSION

The fundamental contribution of this chapter is in providing a reprogrammable structure in developing energy-aware routing for scalable sensor networks (see also

Figs. 13.13 and 13.14). Computational/networking characteristics of INSPIRE include effective link abstraction; sensor service, allowing network protocols to run efficiently on varying power management schemes; power savings greater; simpler code; multiple network protocols that benefit from coexistence, coordination, and cooperation; effective separation of mechanism and policy; building blocks for a sensor network architecture; and potential application to internet architecture and 802.11.

PROBLEMS

13.1 List and explain in your own words four factors that degrade the performance of a wireless sensor network.

13.2 List three ways in which the factors explained above can be improved to extend the life of a sensor network.

13.3 List some advantages that the INSPIRE framework provides to existing sensor networks.

13.4 How much of a performance hit is taken when power-aware routing protocols are employed in sensor networks?

13.5 Compare two (2) other network simulators with GlomoSIM discussed in this chapter.

REFERENCES

1. V. Iyer, S. S. Iyengar, N. Balakrishnan, V. Phoha, and M. B. Srinivas, Farms: Fusionable ambient renewable MACs, *Proc. IEEE Sensors Applications Symp. SAS 2009,* Feb. 17–19, 2009, pp. 169–174.
2. V. Iyer, S. S. Iyengar, G. Rama Murthy, M. B. Srinivas, and B. Hochet, Multi-hop scheduling and local data link aggregation dependent qos in modeling and simulation of power-aware wireless sensor networks, ACM. IWCMC, June 21–24, 2009, Leipzig, Germany.
3. V. Iyer, G. Rama Murthy, M. B. Srinivas, and B. Hochet, C-error simulator for development for sensor and location aware sensing applications, *Proc. 3rd Int. Conf. Sensing Technology, ICST 2008,* Nov. 30–Dec. 3, 2008, pp. 192–199.
4. V. Iyer, G. Rama Murthy, and M. B. Srinivas, Environmental measurement os for a tiny CRF-stack used in wireless network, *Modern Sensing Technol.* (special issue) pp. 72–86, (2008). ISSN 1726-5479 copyright 2006 by IFSA.

14 Modeling Sensor Networks Through Design and Simulation*

Design and programming problems are inherent to the understanding of the technology of sensor networks.

—S. S. Iyengar

Wireless sensor networks have the potential to become significant subsystems of engineering applications. Before relegating important and safety-critical tasks to such subsystems, it is necessary to understand the dynamic behavior of these subsystems in simulation environments. There is an urgent need to develop simulation platforms that are useful for exploring both the networking issues and the distributed computing aspects of wireless sensor networks. Current efforts to simulate wireless sensor networks focus largely on the networking issues. These approaches use well-known network simulation tools that are dificult to extend to explore distributed computing issues.

This chapter presents an architecture of a sensor simulator, and a sensor node that is used in the simulator. This chapter further emphasizes that OMNeT++ is a viable discrete-event simulation framework for studying both the networking aspects and the distributed computing aspects of sensor networks. We present the architecture of a sensor node that is used in the simulator and the general architecture of the simulator.

On the basis of our studies with the IEEE 802.11 MAC and directed diffusion integrated with GEAR, we conclude that our simulator is at least an order of magnitude faster than ns-2 and uses memory more eficiently than ns2. The modular structure of compound modules and the ease of configuring simulation scenarios via an initialization file offers us a tremendous amount of flexibility to model and study the dynamic behaviors of both the sensor network and the application environment in which such networks are expected to operate.

Discrete-event simulation is a trusted platform for modeling and simulating a variety of systems. We discuss results obtained from a simulator for wireless sensor networks that is based on the discrete-event simulation framework called OMNeT++.

* The authors would like to acknowledge the contributions of the former graduate students C. Mallanda, A. Suri, and V. Kunchakarra, and of Drs. R. Kannan, A. Durresi, and S. Sastry. This work was supported in part by NSF-ITR under IIS-0312632 and IIS-0329738 and DARPA/AFRL grant #F30602-02-1-0198.

This framework develops simulations for the 802.11 MAC and the well-known sensor network protocol called *directed diffusion*. The performance of the simulator is deomnstrated by comparing its performance to that of the well-known simulator ns-2. Results indicate that the OMNeT++ simulator executes at least an order of magnitude faster than ns-2 and makes more efficient use of the available memory. The ease of modifying the sensor network and scalability, which is defined as the number of nodes that can be simulated, are two distinguishing features of this simulator.

14.1 INTRODUCTION

Wireless sensor networks (WSN) [5] consist of numerous tiny sensors that are deployed in spatially distributed terrain. These sensors are endowed with a small amount of computing and communication capability and can be deployed in ways that wired sensor systems cannot. For example, sensors can be deployed in environments that are in accessible for humans, or sensor networks can be deployed in environments that are changing such as a chemical cloud. Despite the prolific conceptualization of sensor networks as being useful for large-scale military applications, the reality is that the best migration path for sensor networks research into nonacademic applications is via integration with existing engineering application infrastructure. For example, sensor networks have the potential to offer fresh solutions to fault diagnosis, health monitoring, and innovative human–machine interaction paradigms [4,20,13,19].

Before emerging technologies such as sensor networks and the underlying node-level architectures such as the event-driven architecture of Tiny OS [11] can be incorporated as subsystems in mainstream engineering applications, it is necessary to demonstrate the efficiency and robustness of these subsystems through comprehensive simulations that involve the dynamics of both the application and the sensor network. Such simulation studies must explore the effects of scale, density, node-level architecture, energy efficiency, communication architecture, failure modes at node and communication media levels, system architecture, algorithms, protocols, and configuration, among other issues. Unlike traditional computer systems, it is not sufficient to simulate the behavior of the sensor network in isolation because of the tight and ubiquitous coupling between the sensor network and its application.

14.2 WHY A NEW SIMULATOR

In a recent report [1] the following paragraph summarizes the need for a new simulator.

> ns2, perhaps the most widely used network simulator, has been extended to include some basic facilities to simulate Sensor Networks. However, one of the problems of ns2 is its object-oriented design that introduces much unnecessary interdependency between modules. Such interdependency sometimes makes the addition of new protocol models extremely difficult, only mastered by those who have intimate familiarity with the simulator. Being difficult to extend is not a major problem for simulators targeted at traditional networks, for there the set of popular protocols is relatively small. For

> example, Ethernet is widely used for wired LAN, IEEE 802.11 for wireless LAN, TCP for reliable transmission over unreliable media. For sensor networks, however, the situation is quite different. There are no such dominant protocols or algorithms and there will unlikely be any, because a sensor network is often tailored for a particular application with specific features, and it is unlikely that a single algorithm can always be the optimal one under various circumstances.

Many other publicly available network simulators, such as JavaSims, SSFNet, and Global Mobile Information System Simulator (GloMoSim) and its descendant Quaint, attempted to address problems that were left unsolved by ns2. Among them, JavaSIM developers realized the drawback of object-oriented design and tried to attack this problem by building a component-oriented architecture. However, they chose Java as the simulation language, inevitably sacrificing the efficiency of the simulation. SSFNet and GloMoSim designers were more concerned about parallel simulation, with the latter more focused on wireless networks. They are not superior to ns2 in terms of design and extensibility.

The design of wireless sensor networks requires us to simultaneously consider the effects of several factors such as energy efficiency, fault tolerance, QoS demands, synchronization, scheduling strategies, system topology, communication, and coordination protocols. This chapter presents the structural design of a new simulator for wireless sensor networks that is based on the discrete-event simulation [9,16] framework OMNeT++ and results that demonstrate that the new simulator executes at least an order of magnitude faster than ns2 while using memory more efficiently. While the design we present is general, the simulations focus on an implementation of the IEEE 802.11 MAC layer and directed diffusion integrated with the GEAR (geographic and energy aware routing) protocol.

The remainder of this chapter is organized as follows: Section 14.3 describes the background for simulating sensor networks. Section 14.4 describes the simulation problem. Section 14.5 describes the implementation details of the new simulator, and Section 14.6 discusses the performance of the simulator.

14.3 CURRENTLY AVAILABLE SIMULATORS

ns2 is a well-established discrete-event simulator that provides extensive support for simulating TCP/IP, routing, and multicast protocols over wired and wireless networks [8]. A radio propagation model based on two-ray ground reflection approximation and a shared media model in the physical layer, an IEEE 802.11 MAC protocol in the link layer, and an implementation of dynamic source routing for the network layer were developed in the Monarch project [14].

SensorSIM builds on ns2 and claims to include models for energy and the sensor channel [18]. At each node, energy consumers are said to operate in multiple modes and consume different amounts of energy in each mode. The sensor channel models the dynamic interaction between the physical environment and the sensor nodes. This simulator is no longer being developed and is not available.

OPNET Modeler is a commercial platform for simulating communication networks [17]. Conceptually, the OPNET model comprises processes that are based on finite-state machines, and these processes communicate as specified in the top-level model. The wireless model is based on a pipelined architecture to determine connectivity and propagation among nodes. Users can specify frequency, bandwidth, and power among other characteristics, including antenna gain patterns and terrain models.

J-Sim is another object-oriented, component-based, discrete-event, network simulation framework written in Java [21]. Modules can be added and deleted in a plug-and-play manner, and J-Sims is useful for both network simulation and emulation by incorporating one or more real sensor devices. This framework provides support for target, sensor and sink nodes, sensor channels and wireless communication channels, physical media such as seismic channels, power models, and energy models.

GloMoSim is a collection of library modules, each of which simulated a specific wireless communication protocol in the protocol stack [25]. It is used to simulate ad hoc and mobile wireless networks.

14.3.1 The OMNeT++ Framework

Objective Mocular Network Testbed in C++ (OMNeT++) is a public-domain, component-based, modular simulation framework [23]. It is has been used to simulate communication networks and other distributed systems. The OMNeT++ model is a collection of hierarchically nested modules as shown in Fig. 14.1. The top-level module is also called the *system module or network*. This module contains one or more submodules, each of which could contain other submodules. The modules can be nested to any depth, and hence it is possible to capture complex system models in OMNeT++. Modules are distinguished as being either simple or compound. A simple module is associated with a C++ file that supplies the desired behaviors that encapsulate algorithms. Simple modules form the lowest level of the module hierarchy. Users implement simple modules in C++ using the OMNeT++ simulation class library. Compound modules are aggregates of simple modules and are not directly associated with a C++ file that supplies behaviors. Modules communicate by exchanging messages. Each message may be a complex data structure. Messages may be exchanged directly between simple modules (on the basis of their unique ID)

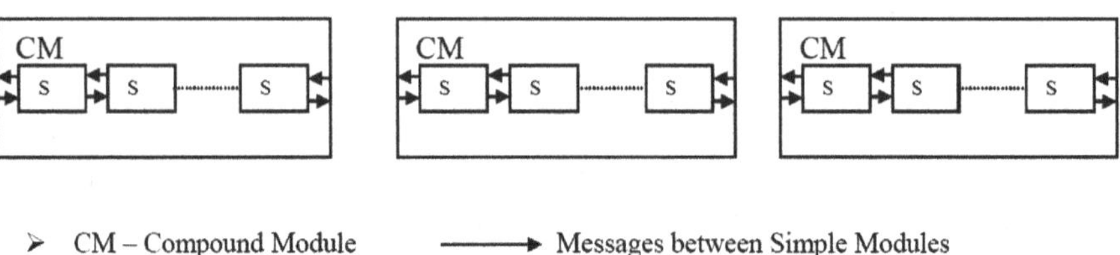

FIGURE 14.1 Simple and compound modules in OMNeT++.

or via a series of gates and connections. Messages represent frames or packets in a computer network. The local simulation time advances when a module receives messages from another module or from itself. Self-messages are used by a module to schedule events at a later time. The structure and interface of the modules are specified using a network description language. They implement the underlying behaviors of simple modules. Simulation executions are easily configured via initialization files. They track the events generated and ensure that messages are delivered to the right modules at the right time.

To take advantage of these features of OMNeT++, we have chosen it as the framework for sensor network simulations. Its salient features include the following:

- OMNeT++ allows the design of modular simulation models, which can be combined and reused flexibly.
- It is possible to compose models with any granular hierarchy.
- The object-oriented approach of OMNeT++ allows the flexible extension of the base classes provided in the simulation kernel.
- Model components are compiled and linked with the simulation library, and with one of the user interface libraries to form an executable program. One user interface library is optimized for command-line and batch-oriented execution, while the other employs a graphical user interface (GUI) that can be used to trace and debug the simulation.
- OMNeT++ offers an extensive simulation library that includes support for input/output, statistics, data collection, graphical presentation of simulation data, random-number generators, and data structures.
- OMNeT++ simulation kernel uses C++, which makes it possible to embed it in larger applications
- OMNeT++ models are built with NEtwork Description (NED) and omnetpp.ini and do not use scripts, which makes it easier for various simulations to be configured.

The following sections give the detailed implementation of our simulation scenario on OMNeT++.

14.4 SIMULATION DESIGN

This section presents the architecture of a sensor node and the overall design of our new simulator [6,10,15]. The topology of the sensor network field in our simulations is derived from the simple and compound module concept of the OMNeT++ framework. As shown in Fig. 14.1, layers of a node behave as simple modules and a sensor node behaves as a compound module, and all these sensor nodes constitute the sensor network depicted as a system module. The architecture of a sensor node is depicted in Fig. 14.2. Each layer of the sensor node is represented a simple module of OMNeT++. The layers communicate with each other through gates, and each layers has a reference to the coordinator. The structure of a layer is represented in Fig. 14.3. These simple modules are connected according to the layered architecture of a sensor

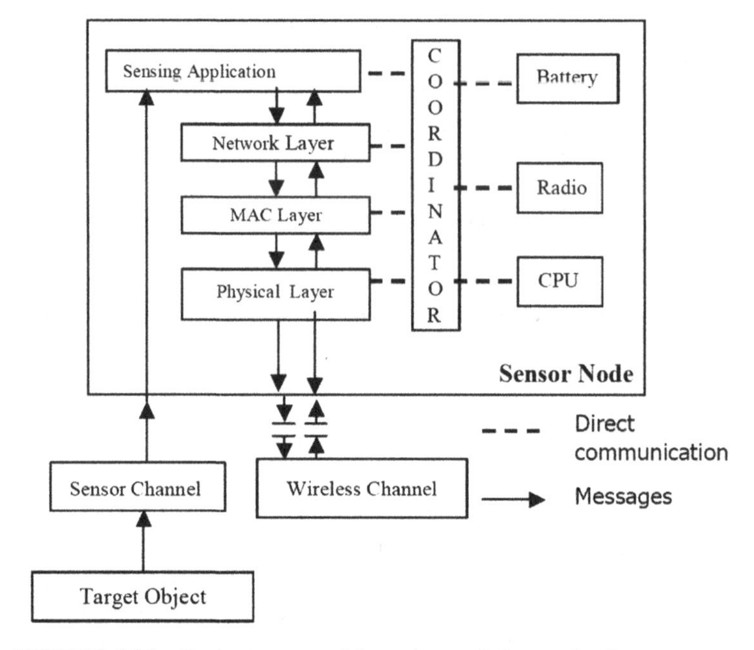

FIGURE 14.2 Basic structure of the sensor node in our simulator structure.

node. The different layers of the sensor node have gates to the other layers of the sensor node to form the sensor node stack. A simple module with wireless channel functionality is used to communicate with these compound modules (sensor nodes) through multiple gates. The functionalities provided by each module are described below with radio, CPU, and battery module forming the hardware model of the sensor node.

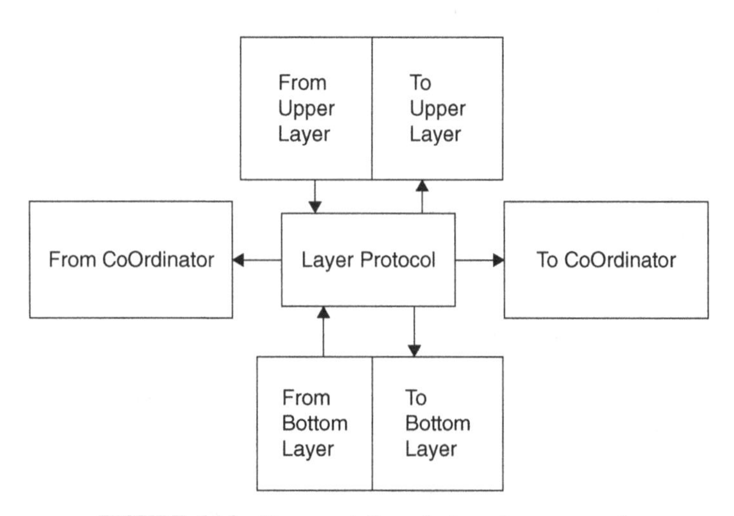

FIGURE 14.3 Representation of a layer in sensor node.

14.4.1 Coordinator Module

The coordinator (formal proprietary name: CoOrdinator) module has the functionalities that coordinate the activities of the hardware and the software modules of the sensor node. It is basically used for interlayer communication. This module needs to be extended, and functionality has to be added for access to properties of new hardware or consumers added. As shown in the Fig. 14.3, the CoOrdinator class has reference to all layers in the sensor node, and all layers in the sensor node may access the CoOrdinator class implementation. Thus, through the coordinator module, any layer may access and update the properties of the other layer. For example, the battery module needs to be informed on transmission or receipt of the packets by the physical module so that the energy consumption is updated at the node accordingly. During simulation the CoOrdinator class is responsible for registering the sensor node to the sensor network. Registering of the sensor node is an indication that the sensor node is up and functioning. When the available energy is completely depleted, the node is deregistered from the sensor network.

14.4.2 Hardware Model

1. *Battery Model.* This module is an essential component of the sensor node, which supplies the necessary energy to the CPU module, radio module, and the sensors used to sense the environment. Hence the battery is connected to all the hardware components of the node and its energy resource decreases depending on the power drawn by all the components. At regular intervals, the module updates its remaining energy depending on the type of battery model used. Various models such as the linear battery model and discharge-rate-dependent model are being implemented. When all the hardware devices report their power consumption, the current discharge of the battery and hence the estimated duration T (in hours), which indicates how long the battery is expected to last, is determined as $T = C/I$, where C is the remaining capacity of the battery in ampere-hours and I is the total current drawn by the sensor node in amperes. The remaining capacity in the battery can be estimated by assuming either a linear model or a discharge-rate-dependent model. In the linear discharge model, the remaining capacity is

$$C = C_{\text{in}} - \int_{\Delta t} I(t)\Delta t$$

where C_{in} is the initial capacity of the battery and $I(t)$ is the current drawn by the sensor node in duration Δt. This model assumes that there are no self-discharges and that the battery does not deteriorate with age. The discharge-rate-dependent model assumes that higher discharge rates effectively reduce the remaining capacity of the battery. To allow various models to be implemented with the type of application, we have a basic battery module, BatteryBase, which forms the abstract class for the different battery models. BatteryLinear is a subclass of BatteryBase and updates the energy depending on the number of consumers and the state of activity of the consumers. BatteryDischargeRate is a subclass of BatteryBase, and the energy consumption is a linear function of current.

2. *CPU Model.* The nodes in a sensor network are usually equipped with very low-end processors or microcontrollers. The power consumption for performing various operations should be very low, and we have used a standard set of parameters for energy consumption by the processor model. The processor needs different levels of energy consumption in the idle, sleep, and active states. The processor power consumption model is very important, and ignoring it will lead to incorrect trends in power consumption in the network. New processor models with enhanced features and improved energy consumption levels can be incorporated in this module for testing various kinds of applications. The CPU Base abstract class forms the basis for different CPU models and defines the interfaces of this module with the coordinator and the battery. CPU Simple implements the power consumption of the CPU in different states: idle, sleep, and active.
3. *Radio Model.* This model is used to characterize the antenna property of a node. RadioBase is an abstract class for the different radio models. RadioSimple, a subclass of RadioBase, updates the energy of the battery depending on the state of the radio: idle, sleep, transmit, and receive. The values for the different properties of the hardware and consumers may be provided through the configuration file.

14.4.3 Wireless Channel Model

The wireless channel module controls and maintains all potential connections between the sensor nodes. These static connections are provided from all the nodes to the wireless channel module and from the module to all the nodes in the NED file. These connections enable sensor nodes to exchange data and communicate with each other. Any message from a node is sent to all the neighbors within its transmission region with a delay d, where d is (distance between communicating sensor nodes)/speed of light.

Various radio propagation models are used to predict the received signal power of each packet. These models affect the communicating region between any two nodes and are derived by the wireless channel.

1. *Free-Space Propagation Model.* The free-space propagation model assumes the ideal propagation condition that there is only one clear line-of-sight path between the transmitter and the receiver. H.T. The received signal power in free space at distance d from the transmitter is estimated as [2]

$$P_r = \frac{P_t * G_t * G_r * \lambda^2}{(4\pi)^2 * d^2 * L^2}$$

where P_t is the transmitted signal power; P_r is the received signal power; G_t, G_r are the antenna gains of the transmitter and the receiver, respectively; L is the system loss; and λ is the wavelength.

2. *Two-Ray Ground Reflection Model.* A single line-of-sight path between two mobile nodes is seldom the only means of propagation. The two-ray ground reflection model considers both the direct path and aground reflection path. This model gives more accurate prediction at a long distance than the free space model. The received power at distance d is predicted by:

$$P_r = \frac{P_t * G_t * G_r * h_t^2 * h_r^2}{d^4 * L}$$

where h_t and h_r are heights of transmit and receive antennas, respectively.

This last equation shows a power loss more rapid than that for the free-space-model as distance increases.

14.4.4 Sensor Node Stack

The simple module at the highest level of the hierarchy of the sensor node, namely, AppLayerSimple, simulates the behavior of the application layer. This module communicates with the NetLayerBase Module through gates to schedule any messages. New applications can be incorporated to this module. The functionality of this module is described in greater detail for the directed diffusion implementation.

The simple network module simulates the packets sent and received by the nodes in the network. The network module initially receives application-layer messages from the AppLayer module and adds the network header to it. The particular features of this layer depend on the protocol implementation. Directed diffusion with GEAR is implemented at the network layer as described in the next section. The packet structure of the network layer sent to the MAC layer has the next hop in the route. The MAC layer provides the interface between the physical layer and the routing layer. It has the basic functionality of media access; the functionality of this module is described in greater detail for the 802.11b implementation.

Such a modular structure of entities simulated with OMNeT++ makes our simulation more flexible than ns2.

14.5 IMPLEMENTATION DETAILS

Using the above mentioned design for the simulator, we implemented directed diffusion at the network layer and compared the performance with the existing simulator ns2. MAC 802.11b is also implemented at the MAC Layer, and the performance is compared with ns2 with directed diffusion at the network layer. The implementation is described with a block diagram in Fig. 14.4. The implementation used a simple pass

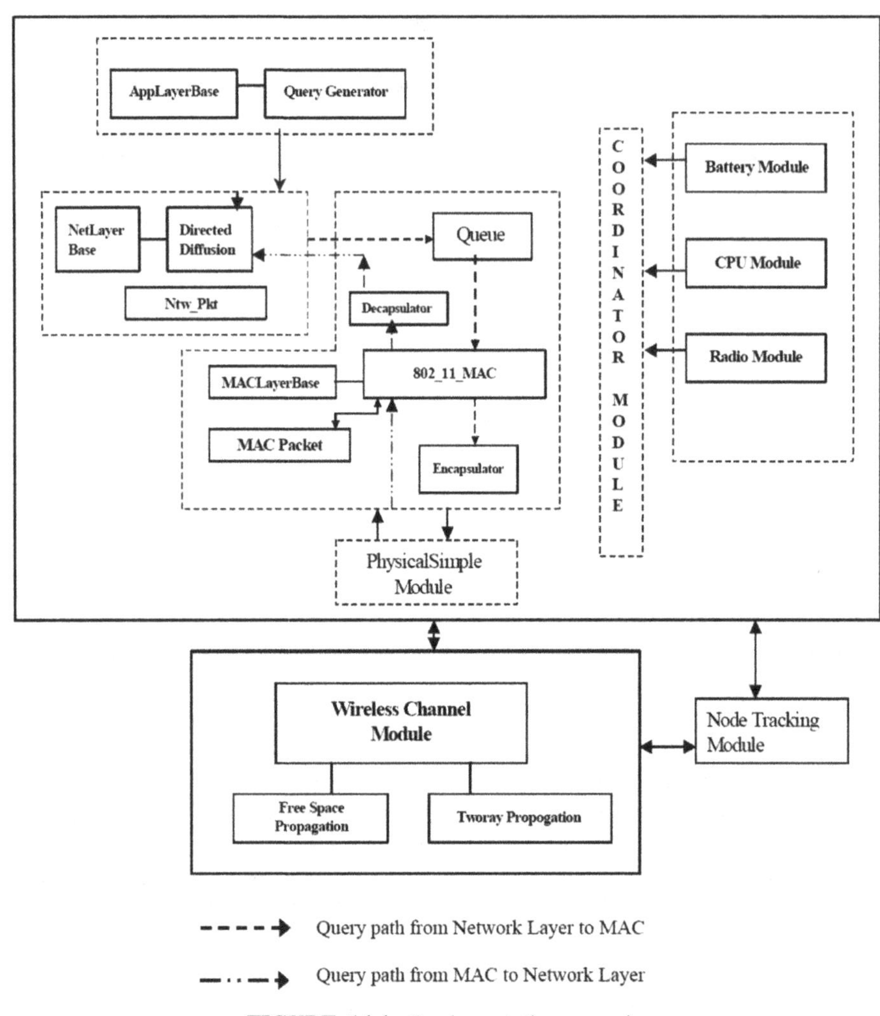

FIGURE 14.4 Implementation scenario.

through the physical layer, a simple wireless channel module with the application layer generating query packets, and forwarding to the network layer.

14.5.1 Directed Diffusion with GEAR

We have implemented directed diffusion [12] along with geographic routing. The application layer generates interests that specify the region, the kind of data required, and rate of delivery of data. Nodes that initiate the interest are called *subscribers*. On receiving the interest message, the network layer broadcasts beacon messages in the network. The immediate neighbors of the node, on receiving beacon messages, reply

Query - attribute
Query - rate of data
Query - duration

FIGURE 14.5 Structure of a query.

back with a beacon-reply-type message that contains their geographic location and the energy left in them. On receiving the beacon reply messages, the neighbor table of the node that sent the beacon is updated. The node waits for a fixed duration of time to receive the beacon reply from all the neighbors. The interest message is then forwarded to the node that has a lower estimated cost to the region as calculated by the GEAR protocol [24]. The next node follows the same procedure and forwards the message toward the region by geographic routing. If a node in the path does not have any neighbors or all its neighbors are away from the region, then it sends a message to its parent node that it is a deadend. The parent node, on updating the cost of the unreachable node, forwards the query (see Fig. 14.5 for query structure) in an alternate route toward the region. In the target region, the interest is disseminated by using recursive flooding. The interest cache is maintained at each node in the path with its gradient of interest to each neighbor. The nodes in the region that have the specified properties of the interest send out data. Nodes that send data out are referred to as *publishers*.

The data are marked as *exploratory* to reinforce the path that was taken by the interest. On receipt of the data marked as *exploratory* by the subscriber, a positive reinforcement message is sent out by the subscriber node. Each node on the path forwards this message, thus reinforcing the path to the region. When a node reinforces a path, its cost to the region is known and this cost is sent back to its source node, which updates the cost information of that node to the particular region of interest. Thus the path with the lowest cost is always maintained, reinforcing the route.

The data from the region follow the path established by the reinforced messages. The nodes in the region send out data at the rate that is specified in the query. Data caching is implemented in intermediate nodes, and so the data requested by different subscribers from the same region can be satisfied by the common node in the path, thus reducing the traffic and redundant messages. The data marked as *exploratory* are sent to identify better paths and reinforce at regular intervals. Also, the neighbor-updating procedure is carried out; specifically, at regular intervals the beacon messages are broadcast and beacon reply messages are sent by neighbors, thus maintaining the latest neighbor information.

14.5.2 802.11 MAC

The MAC layer places the network packet on the wireless channel. The network packet may be a broadcast or unicest packet to a specific node (sink node). Any network-layer packet received by the MAC-802-11 [7,22,3] module is encapsulated into the MAC frame with the MAC header added to it. The network-layer packets have information on whether the packet has to be broadcast or unicast. The broadcast packet is encapsulated into the broadcast MAC frame with the appropriate MAC header and is inserted into the messages queue of the MAC layer. If the network packet is for a particular destination, an RTS frame is created and is inserted in the messages queue of MAC layer. If the network packet length is more than that of the MAC frame, it is fragmented and the fragments for that network packet are created with MAC headers and are inserted into the fragments queue.

The MAC layer then waits for the channel to be idle to send its frame from the messages queue. The MAC layer has a NAV timer, which specifies the busy/idle state of the medium. The NAV timer set for a node implies that the channel is busy. When the NAV timer expires, the MAC layer waits for the channel to be free for DIFS time and if the channel is still idle after DIFS timer has expired, it then goes into exponential backoff. It then waits for a random time set by the backoff timer. The backoff timer decrements its value during the idle period of channel. The node whose backoff timer expires earlier will get the chance to transmit its next frame. All the intermediate nodes receive this frame and set their NAV timers to the values obtained from the header field of the received frame. Then the backoff timer of the intermediate nodes is stopped from decrementing. Once the channel becomes idle (when the NAV timer expires), all the nodes start decrementing their backoff timers. The node whose backoff timer expired earlier and got the channel will send the first message from the messages queue. If it is a broadcast message, then all the nodes in its region receive it and the MAC layers of those nodes encapsulate the network packet and send it to the network layer. If it is a RTS frame, the destination node checks whether its NAV timer is set (whether its transmission region is busy) and then responds to it by sending CTS. All the other intermediate nodes receiving this RTS update their NAV timers to the CTS+DATA+ACK duration, which implies that the channel is busy for that duration, and hence they refrain from transmitting during this interval. If the destination node receives more than two RTS requests within a given time interval, then collision occurs and the destination node does not respond (send CTS) to any of these RTS requests. The source nodes that are sending RTS have an RTSExpired timer set for RTS frames, when they are sent to the destination node. This timer is scheduled to expire after RTS+CTS duration. If the source node does not receive CTS within this duration, RTSExpired timer expireds and the retry counter of that RTS frame is incremented. If the retry counter is less than ShortRetieLimit (as per the specification), then the contention window is doubled and the random time set by the backoff timer is chosen between 1 and the contention window size. If the retry counter reaches ShortRetryLimit, then the message (RTS and corresponding fragment) is dropped by the MAC.

If the destination node responds to RTS by sending back the CTS, the intermediate nodes for CTS will update their NAV timers obtained from the header field of the CTS frame (DATA + ACK duration) and hence refrain from transmitting during this interval. Once the source node gets the CTS, it will send the corresponding fragment of the network packet to the destination and wait for an acknowledgment.

TABLE 14.1 Parameters Used for Our 802.11 MAC Implementation

Property	Value
SIFS[a]	10 μs
DIFS[b]	28 μs
Slot time	20 μs
Data rate	1 Mbps
RTS length	44 bytes
CTS length	38 bytes
ACK length	38 bytes
DATA length	Variable

[a]Short interframe space.
[b]Distributed interframe space.

The destination node, on receiving the data frame, extracts the network packet, sends it to the network layer, and sends back the acknowledgment to the source node. Once the source node gets the acknowledgment, it sends any other fragments that are to be sent to this node without any additional RTS frames. Table 14.1 shows the standard parameters used for our implementation.

14.6 EXPERIMENTAL RESULTS

In this section, we present results from three different studies. First, we establish that the directed diffusion simulation in our work is consistent with the diffusion implementation in ns2. Next, we compare the performance, with respect to execution time and memory used, between our simulation and that of ns2.

14.6.1 Validating Directed Diffusion Implementation

In this experiment, we considered sensor networks with different numbers of nodes between 5 and 200. For each sensor network, we identified the maximum size of the sensor field (with respect to grid coordinates). Then, we identified a fixed number of query-generating nodes and distributed these nodes randomly in the sensor field. Next, we determined a target region and specified the boundary of the region in terms of the grid coordinate and the number of sensor nodes in the region. We then executed the simulation for a specified duration and observed the ratio of the number of packets generated in the region to the number of packets received by the query-generating nodes. The IEEE 802-11 MAC was considered at the MAC layer with a simple pass through the physical layer for these simulations. The results for 5-2000 nodes, shown in Fig. 14.6 for SensorSimulator and ns-2, indicates that for a similar topology and simulation environment, the delivery ratio is comparable with ns2.

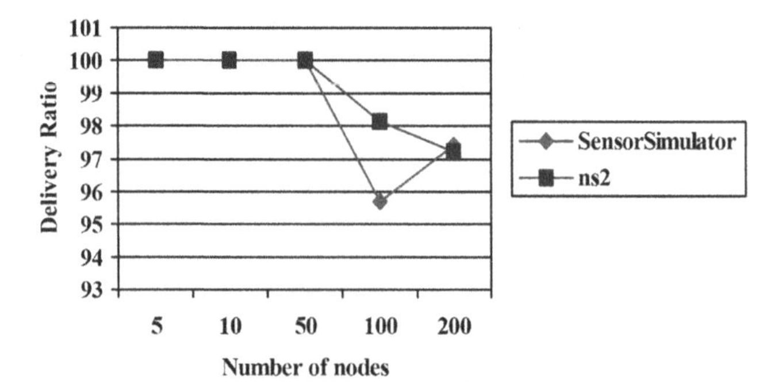

FIGURE 14.6 Comparison of delivery ratio.

We also executed the simulations to verify the directed diffusion implementation on our sensor simulator by observing the changes in delivery ratio of data packets by region, by varying the number of queries as shown in Fig. 14.7.

- Number of nodes = 500
- Simulation time = 300 seconds
- Network size = 500×500 Meter square area
- Number of nodes in region = at least 10 non-faulty nodes per cluster

14.6.2 Directed Diffusion with Simple MAC

Assume that *N* sensor nodes are randomly placed in a grid of size *MP*. Randomly few nodes send queries toward a region of interest. The path taken by queries is decided by first sending interests. We implemented an attribute list to define the type of interest or data message. When a node receives an interest message, it first checks wheather it has the property list of its neighbors. The property list that the node maintains is the distance from the neighboring node to the final destination and the energy levels of the neighboring nodes. If the node has this neighbor list, it checks the last updated time of the neighbor list. If this time is within the permissible limit, this information

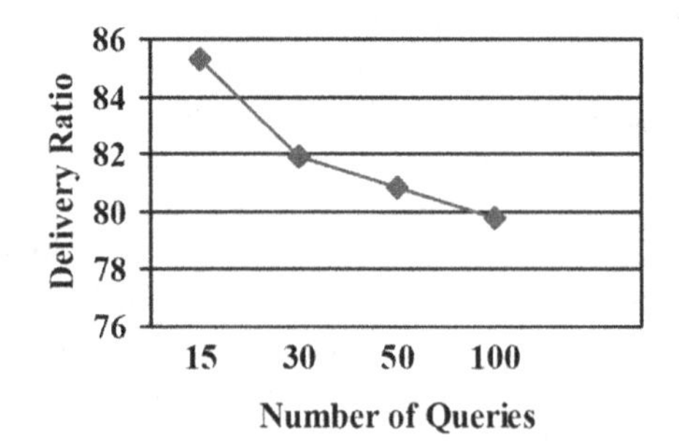

FIGURE 14.7 Delivery ratio of sensor simulator.

is used to decide the next-hop neighbor. If the neighbor list does not exist or the last updated time is more than the desired time limit, then beacon messages are sent out. All the neighboring nodes receive this beacon message. The neighboring nodes then send back beacon reply messages, which update these properties in the neighbor list. The nex-hop neighbor decision is based on the GEAR protocol. We give equal weight age to distance and energy factors. After the query reaches the region of interest, it is flooded to all the nodes in the region. A visited node list is maintained to avoid going into a loop. When a node in the region of interest receives an interest, it sends back an exploratory message to the source of the interest. The exploratory message follows the reverse path taken by the interest message. It gets the reverse path information from the nodes. When this exploratory message reaches the source node, the source node reinforces the path by sending back reinforcements. The reinforcements might or might not follow the same path as the initial interest message. On arrival of the reinforcements, the nodes in the region of interest start sending back data messages at the rate specified in the interest. At regular intervals these data messages are marked as *exploratory*. When the source receives a data message marked as *exploratory*, it sends reinforcements to rebuild the path. This would repair any holes that might have formed in the path.

In order to test the performance of the simulation we ran the setup with queries generated by 10 nodes at random locations in the network. A similar test was performed with 100 nodes generating queries. The queries follow a multihop route to the region following the procedure mentioned above. Once the query reaches the region, the data are sent back once every 5 s for the complete simulation time by all the nodes in the region. The objective of this kind of setup is able to check whether the simulation framework can handle the traffic generated and run to completion as well as to check the amount of time required to run the simulation. Figures 14.8 and 14.9 show the performance of the two simulators (ns2 vs. our simulation) for the setup with 10 nodes and 100 nodes generating queries. For these experiments a pass through simple MAC and a simple physical layer are being considered. It is can be observed that the performance of both the simulators ns2 and SensorSimulator showed similar results with fewer nodes in the network. As the number of nodes in

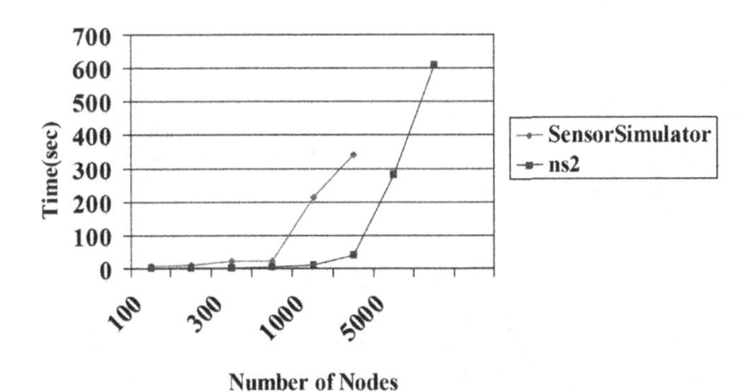

FIGURE 14.8 Execution time for 10 queries: results for SensorSimulator versus ns2.

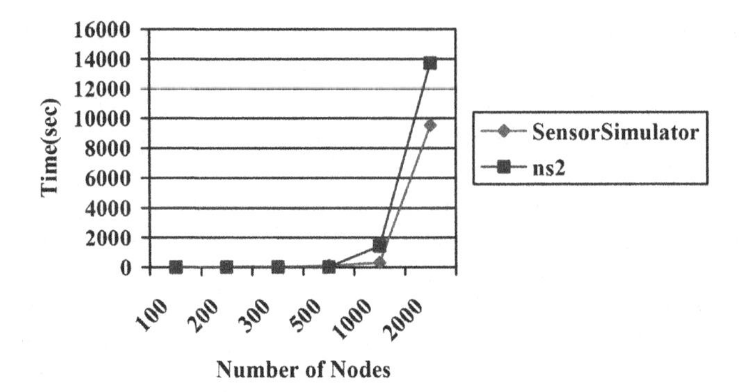

FIGURE 14.9 Execution time for 100 queries: results for SensorSimulator versus ns2.

the network increases, SensorSimulator is able to handle the traffic and the events generated in a better fashion so as to complete the simulation in a reasonable time faster than ns2. It has been observed that ns2 ran out of memory for networks above 2000 nodes. It can be also observed in the figures that the execution time for the simulation run on SensorSimulator is less than that for ns2 for the same simulation results obtained on both the simulators. During the simulation runs, we measured the memory allocated before the start of the simulation, that is, giving the memory usage for the initialization and the setup of the objects of the simulation. The memory usage during the simulation was also measured. The results for the memory usage are as shown in Fig. 14.10 and in Fig. 14.11 for 10 nodes sending queries; Figs. 14.12 and 14.13 show the performance of the simulators for 100 nodes sending queries to the region. This shows that the data structures used for the simulation are used in a scalable manner to represent the different classes and the interaction with the framework. It can also be observed that the rate of memory usage increases at a rate that is, faster for ns2 than for SensorSimulator, thus allowing for large simulation setup and more scalability in SensorSimulator than in ns2.

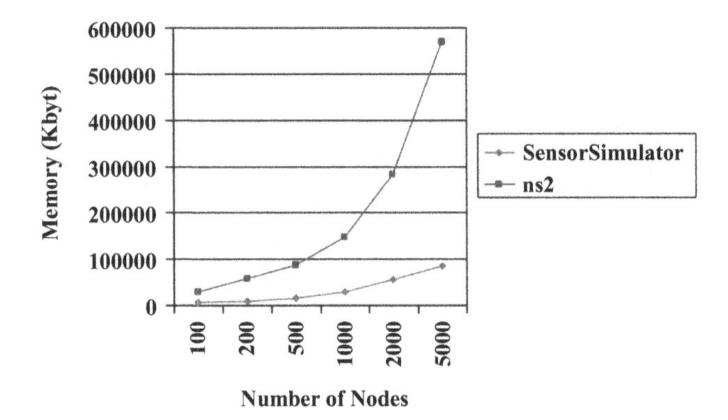

FIGURE 14.10 Memory consumption before simulation starts: 10 queries.

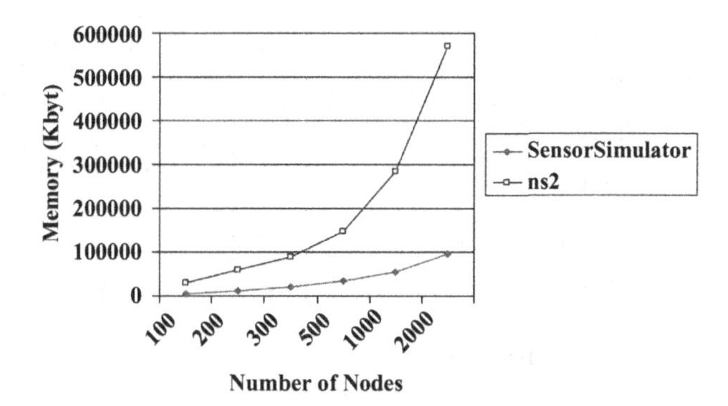

FIGURE 14.11 Memory consumption after simulation ends: 10 queries.

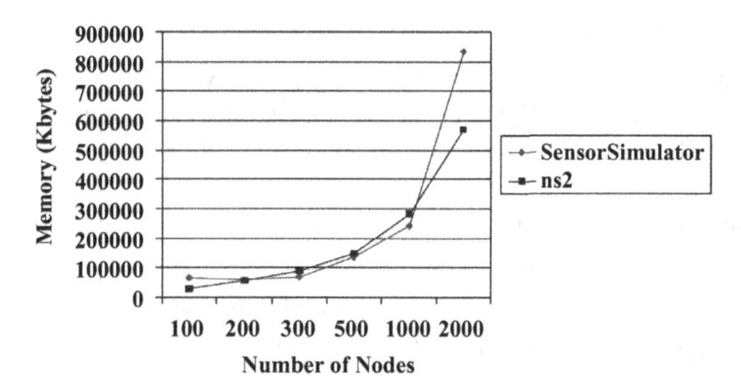

FIGURE 14.12 Memory consumption before simulation starts: 100 queries.

14.6.3 Directed Diffusion with IEEE 802.11 MAC

This series of experiments use MAC 802.11b at the MAC layer and compare the performance of sensor simulator with that of ns2. A simple pass through the physical layer is considered. The simulation is run for 100, 500, 1000, and 2000 nodes. The

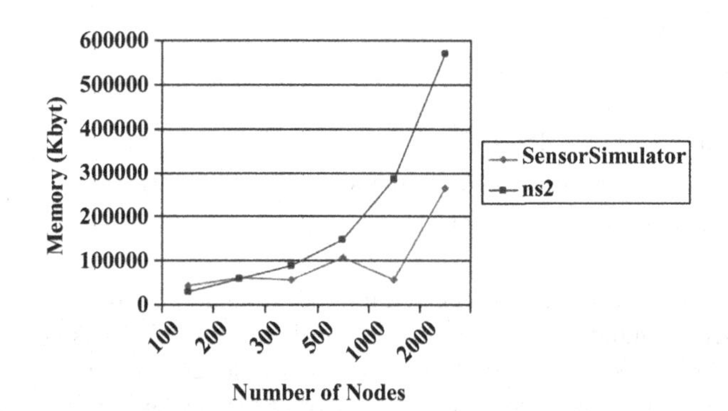

FIGURE 14.13 Memory consumption after simulation ends: 100 queries.

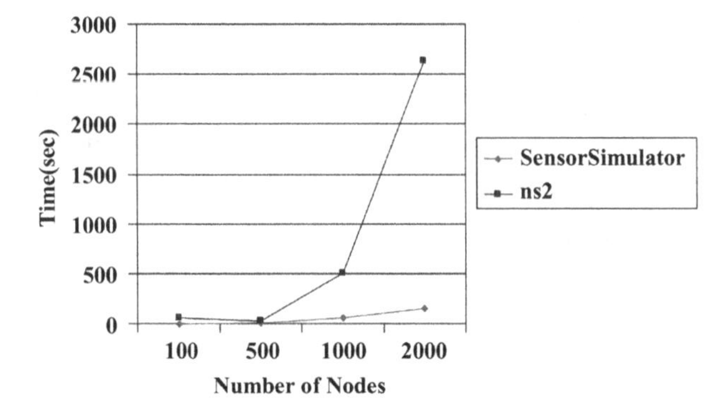

FIGURE 14.14 Comparison with ns2 execution time.

nodes in the sensor network are deployed randomly in various locations with the network size being configurable in the omnetpp.ini file. Figure 14.14 shows the relative performance. The setup for the nodes is as follows:

- Number of queries: 10
- Simulation time: 300 s
- Network dimension: varies with the number of nodes
- Number of nodes in region: 10

A similar scenario was developed in ns2, and the performance of the simulation, that is, the time taken by the simulator to complete the application, was compared in both of them. The results show that SensorSimulator takes less time than ns2 even when the numbers of nodes are increased to 2000. The results were validated by confirming that the query nodes are getting back the appropriate data from the region. The next simulations are carried out for high-traffic scenarios. The number of nodes are varied from 500 to 2000 with 100 nodes generating queries at random intervals. This result, as seen in Fig. 14.15, shows that SensorSimulator is able to perform better than ns2 even for high-traffic networks.

- Number of queries: 100
- Simulation time: 250 s
- Network dimension: varies with the number of nodes
- Number of nodes in region: 15

The memory used was also compared for both simulators, and our observations show that SensorSimulator consumes less space than does ns2. These results, presented in Fig. 14.16, show that the data structures employed for the simulation are used in a scalable manner to represent the different classes and the interaction with the OMNeT++ framework. It can also be observed that the rate of memory usage

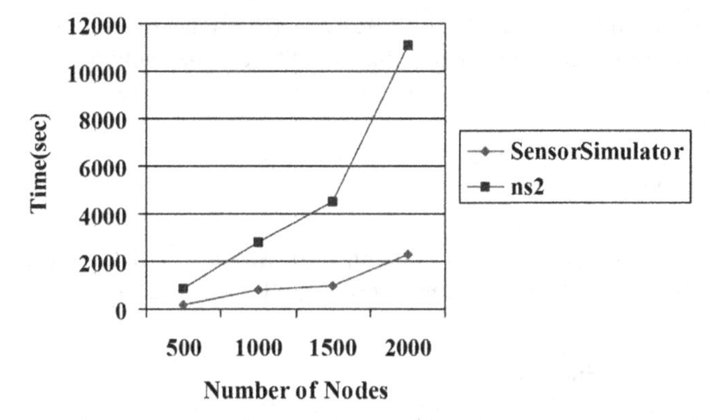

FIGURE 14.15 Comparison of execution time with 100 queries.

increases at a faster rate for ns2 than for SensorSimulator, thus allowing for large simulation setup and more scalability in SensorSimulator than in ns2.

- Number of queries: 10
- Simulation time: 300 s
- Network dimension: varies with the number of nodes
- Number of nodes in region: 5

14.7 FINAL COMMENTS

On the basis of our studies with the IEEE 802.11 MAC and the Directed Diffusion model integrated with GEAR, we conclude that our simulator is at least an order of magnitude faster than ns2 and uses memory more efficiently than does ns2. The modular structure of compound modules and the ease of configuring simulation

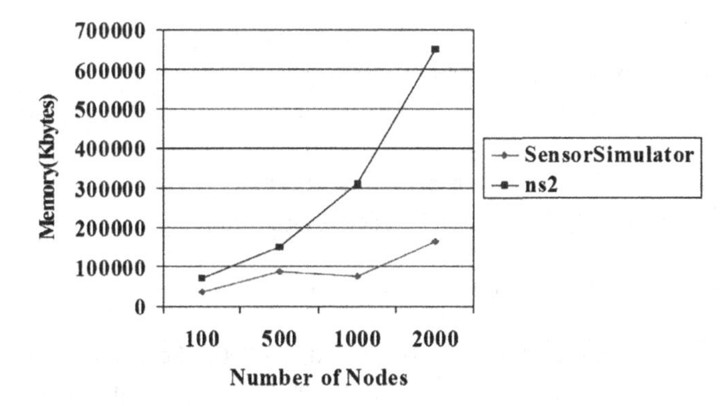

FIGURE 14.16 Comparison of memory consumption.

scenarios via an initialization file offers us a tremendous amount of flexibility to model and study the dynamic behaviors of both the sensor network and the application environment in which such networks are expected to operate.

APPENDIX

This section provides practical guidelines for SensorSimulator software, the source code for which can be found at http://csc.lsu.edu/sensor web/.

What is SensorSimulator?

As discussed in the main text of this chapter, SensorSimulator is a framework developed on OMNeT++, intended mainly to support sensor network simulations. The framework provides basic modules that can be derived in order to implement the users' own modules. With this concept programmers can easily develop their own protocol implementations for the SensorSimulator framework without having to deal with the necessary interface and interoperability issues.

Overview of Framework

The section gives the basic framework of SensorSimulator on OMNeT++. The user needs to know about programming in OMNeT++. If not, you should read the OMNeT++ manual (available at http://www.omnetpp.org/). The manual then explains how to install SensorSimulator on OMNeT++.

Installation You need a running OMNeT++ version 2.3 to use the LSU SensorSimulator with all of its functionality. This is available from the omnetpp.org Website. The user needs to go through the user manual of OMNeT++ to run simulations on our simulator. After downloading the most recent version of the SensorSimulator from the download area of the LSU Website, copy the file to the desired directory. cd to this directory and then perform the following steps:

1. Extract the `LSU-SensorSimulator.tgz` file. It creates a `LSU-SensorSimulator` folder.
2. Append the path `-LSU-SensorSimulator/src` to `$LD LIBRARY PATH`;
3. Run "make" in LSU-SensorSimulator/samples.
4. You can run the simulation by executing ./Simulation.

Directory Structure Our directory structure is divided as follows:

`LSU-SensorSimulator/inc/` This has the header files of base classes src/. This has the base classes for all the layers: CoordinatorBase, LayerBase, PhyLayerBase, MacLayerBase, and NetLayerBase.

AppLayerBase
RadioBase
BatteryBase
CPUBase
WirelessChannelBase
TargetNodeBase
cConsumer

`samples/` This has the simple classes derived from base classes; it also includes implementation of directed diffusion with GEAR and MAC 802.11:

```
phy_layer/ mac_layer/ netw_layer/ app_layer/
hwmodels_layer/  wireless_ch/ common/
```

Each subdirectory has implementations of that layer.

- Hardware models include battery, CPU, and radio modules. These include simple hardware models.
- The common directory has CoOrdinator, packet structures for network and MAC layers and other constants and attributes used for simulation. This directory is derived by all other directories of the sample folder.
- The wireless channel subdirectory has a simple wireless channel that introduces a delay.

Delay D is calculated as follows:

$$D = \frac{\text{distance between 2 communicating nodes}}{\text{speed of light}}$$

The class name has to be specified in the omnetpp.ini file if the user wants to try the other implementations specified above.

To try a new implementation, the following steps are recommended:

1. *Files*. Each layer has three basic files: .cc, .hand, and .nedfiles. The user needs all three of these files for the implementation. We are providing these three files for every layer with minimum functionality. You can add your own code and just do "make" and run the simulation by changing the necessary configurable parameters in the omnetpp.ini file in the samples directory.
2. *Procedure*. Move into samples directory. This directory has subdirectories for each layer. Each directory has examples of various implementations at each layer. Also, you can see the files `New* Layer.cc`, `New*Layer.h`, and `New*LayerModuleDefined.ned` for a new user to start using the simulator. You can add the various parameters of a module that you will be using in your implementation in the .need file. After adding the code or making changes

for the existing implementation, do make and then go back to the samples directory. The class name of this new implementation has to be specified in omnetpp.ini so that the simulation considers your implementation for execution. (e.g., `sensorNetwork.Nodes[*].strMACLayerType= "MAC 802 11"`) With this functionality, the user can just add his/her protocol at that particular layer with all other protocol layers being the same.

3. *Building Simulations.* This section explains the basic concepts behind the SensorSimulator. The class hierarchy is explained and all relevant functions of the base modules are introduced. Detailed description is also available in the API reference.
4. *Running the Simulation.* We can run the simulation after all the layers are defined by configuring the parameters in the omnetpp.ini file. After writing your code at various layers, do "make" in that corresponding subdirectory. After "make" is done successfully, return to the samples directory and change the parameters in the omentpp.ini file. The topology, simulation time, and other properties can be varied in this file. These sections are clearly explained in the OMNeT++ manual. Change the class name of the module you created (as discussed in Section 14.3) and run the simulation. You can redirect the out put to another text file.

Base-Layer Concept. The functionality of any layer of a node is defined in the Layer Base file. This itself gets its properties from the SimpleModule of OMNeT++. All the base classes of the node are derived from LayerBase. This is defined in the .src directory.

Base Modules. We provide base modules for each layer, which inherits its properties from Layer Base. It is implemented primarily to clearly define the interface that can be understood easily and that can be extended if necessary. It implies that these source code files handle the basic functionalities of that layer and the user need not go through all this for the implementation. Users only need to derive their class from these base modules. These modules have header files in the.inc directory. The .ned files and the source code files are available in the src directory.

Simple Modules. A simple functionality of each layer is defined here. These are derived from base modules. Users must implement their protocols at this level. These are available in the samples directory. The simple source code files for each layer are available in each subdirectory of the samples directory with respect to that layer. The .ned files for these simple source files are available in the src directory. The .ned files for new implementations are available in the subdirectories itself (see item 4 in list preceding "Base-Layer Concept" heading).

Conclusion

We conducted various experiments to verify the data received by the subscribers from publishers. This delivery ratio is 100% for smaller networks. The delivery ratio decreases for larger networks according to directed diffusion.

The user can run directed diffusion with 802.11 by including the class filename in the omnetpp.ini file. The topology, number of queries, number of nodes in the region, the region size, and the simulation time are configurable in the omnetpp.ini. File. The user can also run various simulations using different seeds as described in the OMNeT++ manual.

ACKNOWLEDGMENTS

The authors acknowledge the assistance of students and faculty in the Sensor Network Group for their contributions toward the development of SensorSimulator. The authors also acknowledge the CCT for providing infrastructure facilities for the development of the Sensor Network Laboratory.

PROBLEMS

14.1 Provide a two-page essay on how much realism can be achieved in a network simulator.

14.2 Compare the features of J-Sim, GloMoSim, and OMNet++.

14.3 Why are traditional methods of network simulation inadequate for sensor network simulation?

14.4 Provide a brief write up describing the distinguishing features of the network simulator discussed in this chapter.

REFERENCES

1. http://www.cs.rpi.edu/cheng3/sense/.
2. http://www.isi.edu/nsnam/ns/ns-documentation.html.
3. Anonymous, Information Technology-Telecommunication and Information Exchange between Systems-Local and Metropolitan Area Networks Specific Requirements—Part 11: Wireless LAN Medium Access Control (MAC) and Physical Layer (PHY) Specifications, Technical Report, IEEE Standard, 1999.
4. J. Agre, L. Clare, and S. Sastry, A taxonomy for distributed real-time control systems, *Adv. Comput.* **49**:303–352 (1999).
5. I. F. Akyildiz, W. Su, Y. Sankarasubramaniam, and E. Cayirci, Wireless sensor networks: A survey, *Comput. Networks* **38**:393–422 (2002).
6. S. Basavaraju, *Sensim: A Wireless Sensor Network Simulation Template*, M.S. Project, Dept. Computer Science, Louisiana State University, Baton Rouge.
7. V. Bharghavan, A. Demers, S. Shenker, and L. Zhang, Macaw: Amedia access protocol for wireless lans, *InACM SIGCOMM 1994*, 1994.
8. K. Fall and K. Varadhan, *Ns-2 Network Simulator*, Technical Report, Univ. California, Berkeley, 2004.
9. G. S. Fishman, *Principles of Discrete-Event Simulation*, Wiley, 1978.

10. LSU Research Group, LSU SensorSimulator (LSU Sensim, version 1, Jan. 2005) User Manual, Dept. Computer Science, Louisiana State University, Baton Rouge.
11. J. Hill, R. Szewczyk, A. Woo, S. Hollar, D. Culler, and K. Pister, System architecture directions for networked sensors, *ACM Sigplan Notices* **35**:93–104 (2000).
12. C. Intanagonwiwat, R. Govindan, D. Estrin, J. Heidemann, and F. Silva, Directed diffusion for wireless sensor networking, *IEEE/ACM Trans. Networking* **11**(1):216 (Feb. 2003).
13. S. S. Iyengar and R. R. Brooks, eds., *Distributed Sensor Networks*, CRC Press Dec. 2004.
14. D. B. Johnson, *The Rice University Monarch Project*, Technical Report, Rice Univ., 2004.
15. C. Mallanda, *SensorSimulator: A Simulation Framework for Sensor Networks*, Master's thesis, Dept. Computer Science, Louisiana State University, Baton Rouge.
16. J. Misra, Distributed discrete-event simulation, *ACM Comput. Surveys* **18**(1):39–65 (March 1986).
17. OPNET Technologies Inc., *Opnet Modeler*. www.opnet.com
18. S. Park, A. Savvides, and M. B. Srivastava, Sensorsim: A simulation framework for sensor networks, *Proc. 3rd ACM Int. DRAFT Workshop on Modeling, Analysis and Simulation of Wireless and Mobile Systems*, 2000, pp. 104–111.
19. S. Sastry, Smart space for automation, *Assembly Autom.* **24**(2):201–209 (2004).
20. S. Sastry, S. S. Iyengar, and N. Balakrishnan, Sensor technologies for future automation systems, *Sensor Lett.* **2**(1):9–17 (2004).
21. A. Sobieh and J. C. Hou, A similation Framework for Sensor Networks in j-sim, Technical Report UIUCDCS-R2003-2386, Dept. Computer Science, Univ. Illinois, Urbana–Champaign, Nov. 2003.
22. A. Tannenbaum, *Computer Networks*, Prentice-Hall, 2002.
23. A. Vargas, *Omnet++ Discrete Event Simulation System*, version 2.3, 2003.
24. Y. Yu, R. Govindan, and D. Estrin, *Geographial and Energy Aware Routing: A Recursive Data Dissemination Protocol for Wireless Sensor Networks*, Technical Report, Aug. 2001.
25. X. Zeng, R. Bagrodia, and M. Gerla, Glomosim: A library for parallel simulation of large-scale wireless networks, *Proc. Workshop on Parallel and Distributed Simulation*, 1998, pp. 154–161.

15 MATLAB Simulation of Airport Baggage-Handling System

If we want users to like our software, we should design it to behave like a likeable person.

—Alan Cooper

15.1 INTRODUCTION

This section discusses a simple implementation of an intelligent baggage-handling system in MATLAB. Airport baggage-handling systems (BHSs) fall under the broad umbrella of material-handling systems that automate the process of moving materials. Current BHS controllers are based on standard industrial techniques, namely, Programmable Logic Controllers (PLCs). Distributed techniques consist of either separate but interdependent parts or autonomous agents that can share information. IEC 61499 is a standard for distributed systems.

15.2 BACKGROUND

Sensing devices are central to the functioning of airport baggage-handling systems (BHSs). Almost all airport BHSs employ the following sensors/sensing devices:

- Radar/transponder
- Camera
- Smart cards
- Radiofrequency identification (RFID)
- Motion sensor
- Automatic target recognition (ATR)
- X-ray scanner

Hence, the sensor network is heterogeneous. The area of data fusion deals with combining data from different kinds of sensors. There are two approaches to processing

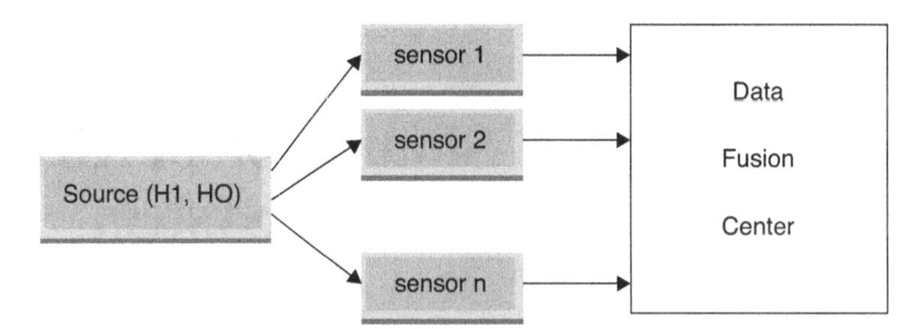

FIGURE 15.1 Flowchart showing sensors processing data.

signals when employing multiple sensors [1]. In the first case, the raw data can be transmitted to a central processor. This approach requires low-latency transmissions and considerably high bandwidth. In the second case, some or the entire signal processing is performed by sensors. This approach addresses latency and bandwidth issues encountered in the first case. In Fig. 15.1 the sensors process the data and relay these intermediate results to the data fusion center, which in turn processes the incoming information. The global results are finally available at the data fusion center. In energy- and processing-rich environment like an airport, the first approach is appropriate.

The baggage is checked in at any of the desks labeled desk 1, desk 2, and desk 3 in Fig. 15.2. During checkin RFID tags are attached to them. After checkin, they are placed on the respective conveyer belts and introduced into the BHS. The scanner that is in close proximity to the location where the bags are introduced reads the RFID tags attached to the bag and energizes the corresponding diverter to feed it to the input of an X-ray scanning machine. If the X-ray scanning machines are currently occupied, the bag's RFID will be saved in a local queue, and the bag will have to follow the loop until it is scanned again by the abovementioned scanner. After the X-ray machine, the state of the bag will be any one of the following:

- The bag passed X-ray scan and deemed eligible for transportation to the aircraft.
- The bag was identified for further manual inspection (level 4 inspection).
- The bag was not able to make its way through the X-ray machine or it was not recorded.

The other scanner placed on the loop with the help of diverters dispatches bags for further inspection or diverts them to a channel that leads to the aircraft. The other scanner placed on the loop with the help of diverters dispatches bags for further inspection or diverts them to a channel that leads to the aircraft [2].

Leone et al. emphasize the importance of relying on computer-based models to simulate the impact of integrating explosive detection devices (at checkin terminals in airports) on the overall operation. To better understand the influence of the checked-baggage screening (CBS) process on the overall operation, a five-stage model is

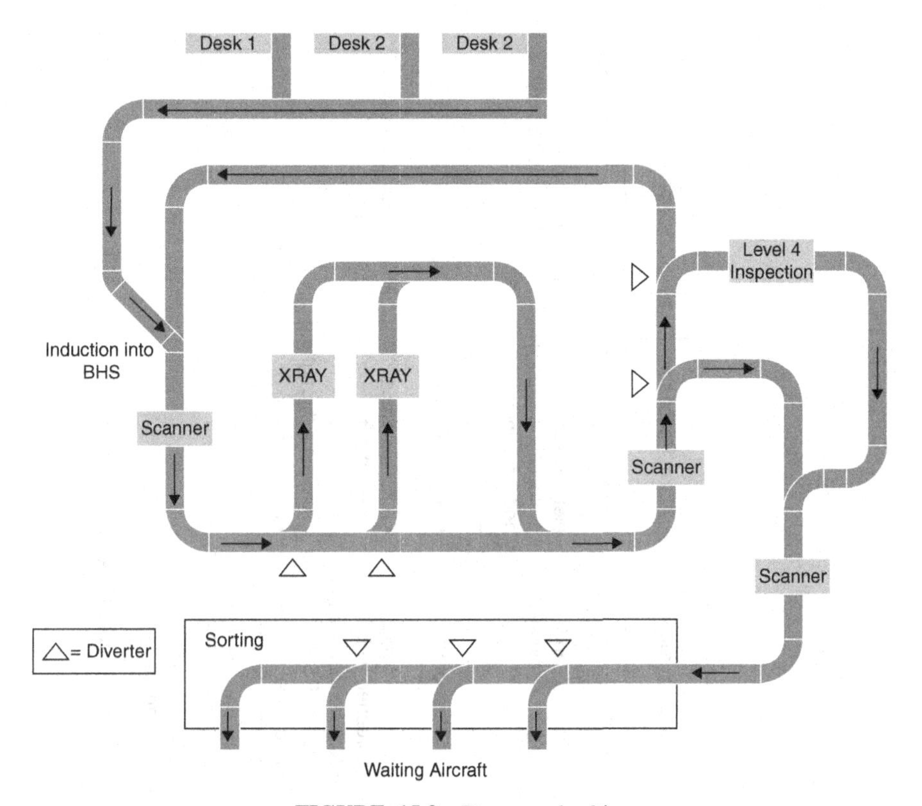

FIGURE 15.2 Baggage checkin.

proposed as shown in Fig. 15.3. This logical view along with other parameters that govern operation (e.g., ticket counter checkin processing time) will serve as input in developing a model to study demand and capacity requirements [4].

Le et al. are concerned mainly with finding the optimal input conditions to ensure optimal throughput performance. Their paper presents an empirical study of a multiobjective problem within a BHS with a goal of estimating the near-optimal input conditions, such that no blockage occurs at checkin stations, while minimizing the baggage travel time and maximizing the throughput performance measures. They provide a practical hybrid simulation and binary search technique for determining a near-optimal-input throughput operating conditions [3]. Our work can be extended to provide a type of simulation similar to this one.

The process of transporting baggage between the aircraft and the terminal that is currently employed is labor-intensive and inefficient [5]. The current transportation scheme is shown in Fig. 15.4. The authors propose an architecture that addresses slow baggage-processing speed, congestion on airport airside roads, inflexibility, and other problems. The proposed architecture is shown in Fig. 15.5. The baggage truck shown in the proposed architecture facilitates temporary storage and automated transfer at the terminal and the aircraft.

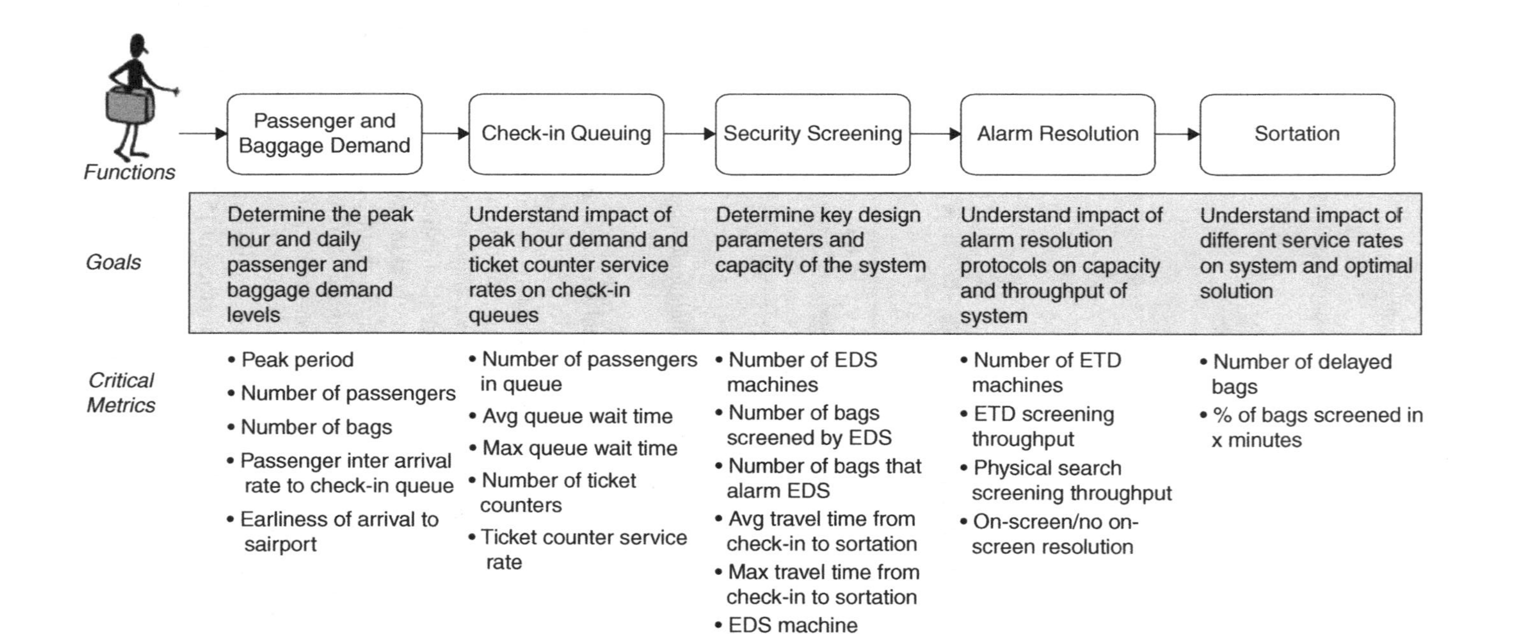

FIGURE 15.3 Screening of checked baggage.

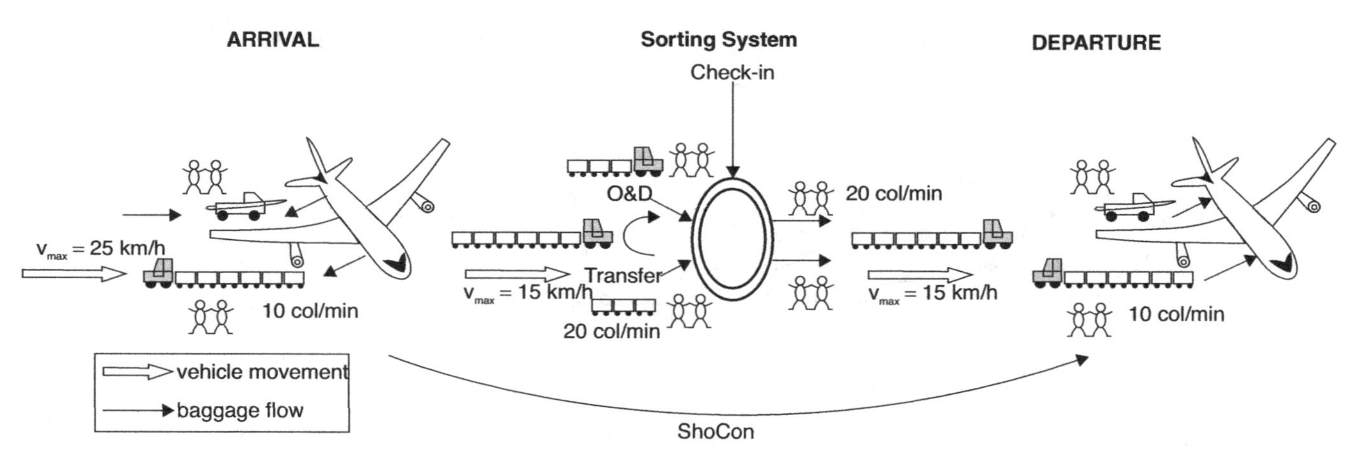

FIGURE 15.4 Baggage transport between airport terminal and aircraft.

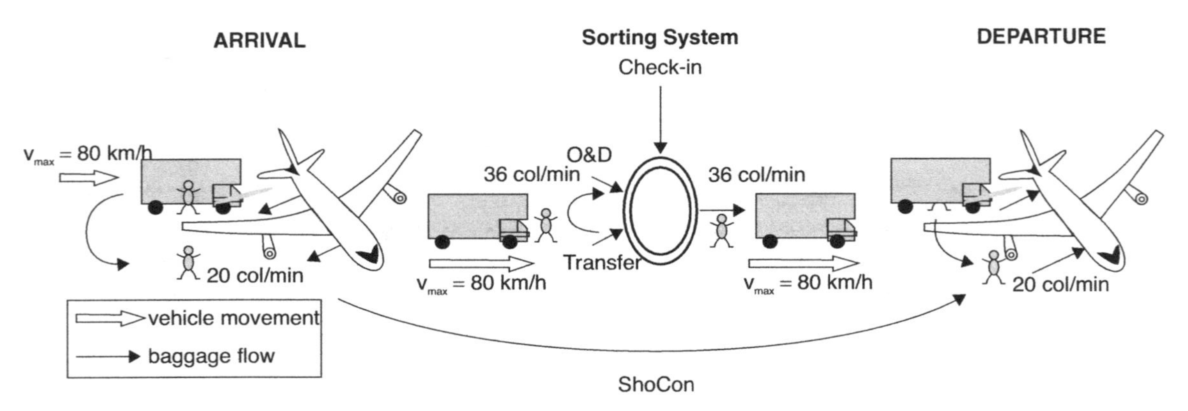

FIGURE 15.5 Proposed architecture for expediting the baggage-handling process.

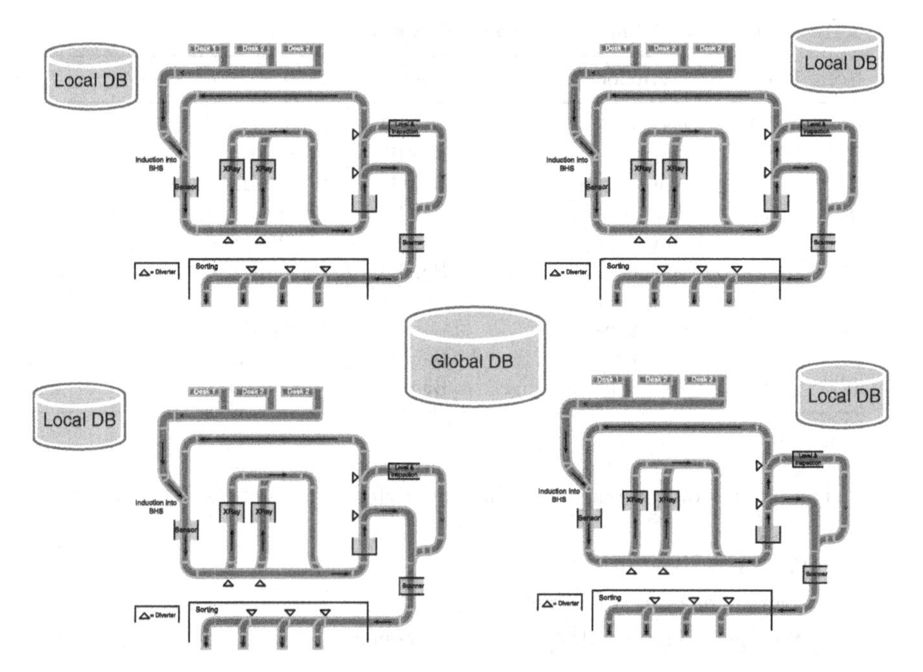

FIGURE 15.6 Detailed flowchart of BHS architecture shown in Fig. 15.5.

15.3 PROPOSED ARCHITECTURE

After reviewing the literature we have adopted the architecture shown in Fig. 15.6. The architecture shows local databases that store relevant transactions at the airport. Data are transferred to and from the global database. For example, when a fight leaves an airport, the pertinent information in the local database is transferred to the global database.

15.4 SIMULATION RESULTS AND DISCUSSION

Our simulation consists of a preexisting data file as shown in Table 15.1. The data file represents data collected from an array of virtual sensors.

The simulation engine is shown in Fig. 15.7. An event handler is required to respond to these events and to firsthand visualization. When an event occurs, control is transferred to the event handler. The event handler executes an order and a message is printed on the console.

Rough pseudocode for the event handler's internal logic is shown in Table 15.2. If the event needs to be visualized firsthand, a procedure is invoked. The visualization currently displays only sensor events that occur at the current timestamp. The MATLAB code for our simulation can be found in Section 15.5. A snapshot of visualization is shown in Fig. 15.8.

TABLE 15.1 Preexisting Data File in BHS Simulation

Event Type	Description	Col2	Col4	Col5
01	Checkin	Bag ID	Flight ID	—
02	Weighin	Bag ID	Weight (lb)	—
03	Security scan	Bag ID	0/1 (success/fail)	—
04	In/out of storage	Bag ID	Airport/storage area ID	—
05	In transfer van	Bag ID	Van ID	—
06	In plane	Bag ID	Flight ID	—
07	Flight departs	—	Flight ID	—
08	Flight arrives	—	Flight ID	—
09	Checkout	Bag ID	Airport ID	Carrier
10	In/out of lost baggage	Bag ID	Airport ID	Carrier

Our simulation computes the following performance metrics of the BHS:

- Percentage bags that missed fiight
- Average bag time to storage
- Average bag time on loop (i.e., time in queue)

The performance metrics obtained for the test data file are shown in Table 15.3.

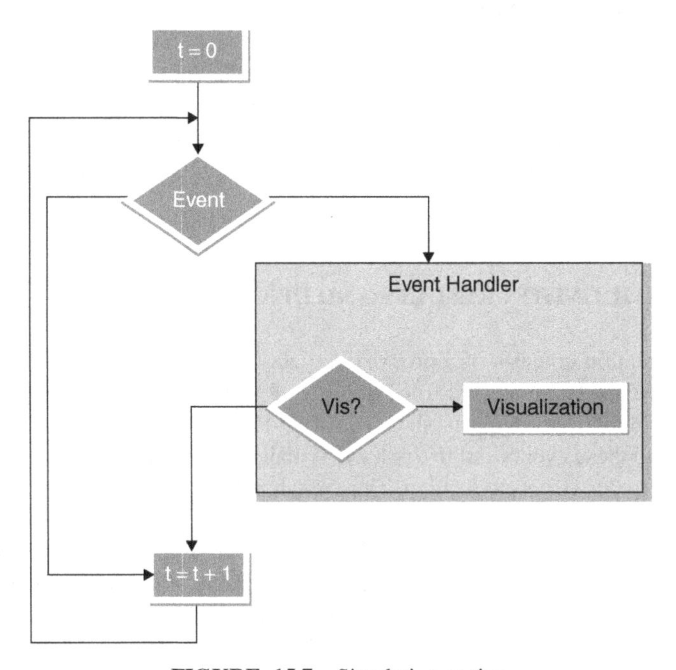

FIGURE 15.7 Simulation engine.

TABLE 15.2 Pseudocode for Event Handler's Internal Logic

Event	Description	Query	Action
01	Checkin	Always true	Photograph person who arrived with bag
02	Weighin	`localdata(4) >Threshold`	Alert person at desk
03	Security scan	`field(4)!=0`	Redirect bag to level 4 inspection
04	Bag. went into storage	`#results(bag==localdata(2) && data type==04) is odd`	Set state to "in storage"
04	Bag. went out of storage	`#results(bag==localdata(2) && data type==04) is even`	Set state to "not in storage"
05	In transfer van	—	—
06	In plane	`#results(bag==localdata(2) && data type==03 && field4==0)==0`	RED ALERT! Unauthorized luggage!
07	Flight departs	`#results((field4==localdata(4)) && field3==01) - (field4==localdata(4)) && field3==06))>0`	Pass results of query to baggage guy, and log them somewhere; set state of bag to "lost"
07	Flight departs	`(field4==localdata(4)) && field3==06) (field(2)==any_RESULT(2))`	Cut and paste results of query to global database; cut and paste local data to database
08	Flight arrives	`(field4==localdata(4)) && field3==06) field(2)==any value in RESULT(2))`	Cut and paste results of query to local database
09	Checkout	True	Take photo
09	Checkout	`(field2==localdata(2))`	Cut and paste results of query to global database
10	In/out of lost baggage	—	—

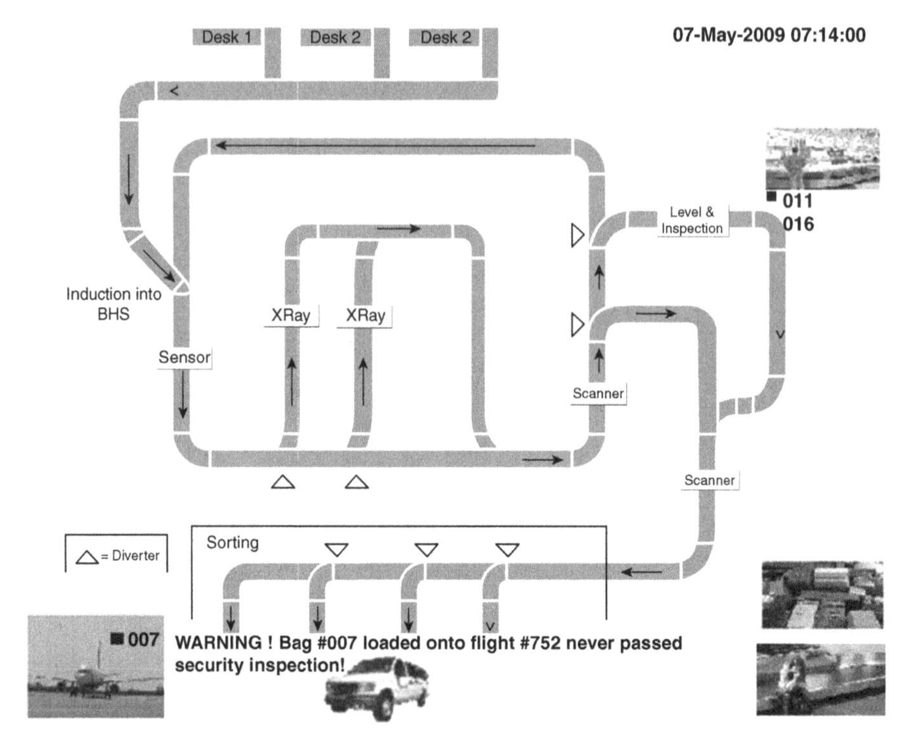

FIGURE 15.8 BHS flowchart showing potential security breach.

15.5 SOURCE CODE

The MATLAB source code for the baggage-handling system is shown below. This program describes a unique way to detect and respond to various combination of events and conditions. Synthetic input data could be generated by hand or by another program (they were generated by hand for testing/metrics). The BHS structure consists of a main loop, an event-handling layer, and an display layer. The main loop continually scans for events and passes them to the event handler, as well as being responsible for computing matrics. The event handler contains the internal logic to actually generate and handle events on the basis of the input data. The display layer is responsible for displaying current events visually.

TABLE 15.3 Performance Metrics for Test Data File

Percentage bags that missed flight	11.1111
Average bag time to storage	15.3333 min
Average bag time on loop	9.5000 min

snproj.m

```
clear all;
close all; clc;

global time
global tmp2

global bags_checked_in_for_flights
global bags_missed_flights
global times_from_check_in_to_storage
global times_waiting_for_scan

global weightthreshold
weightthreshold=50;

A=textread ('final_log.txt', '%s');
startdate=date;
startdate=datenum(startdate);
time=startdate;
currentline=1;
tmp=datevec (getdata (currentline,1,A));
tmp2=datevec(date);
tmp2=tmp2 (1:3);
bags_checked_in_for_flights=0;
bags_missed_flights=0;
times_from_check_into_storage=[ ];
times_waiting_for_scan=[ ];

while ~strcmp(getdata (currentline,1,A), '\0\0')
%while time<startdate+1
      disp (datestr (time));
      vis_was_called=0;
      if ((datenum([tmp2 tmp(4:6)])<=time && ~strcmp (getdata ...
      (currentline,1 ,A ), '\0\0')))
          sensor_visualization_clear ();
          vis_was_called=1;
      end
      while (datenum ([tmp2 tmp (4:6)]) <= time && ~strcmp (getdata ...
      (currentline,1 ,A), '\0\0'))
          %disp (['handle ('num2str (currentline)')']);
          handle_sensor_event (currentline ,A);
          %truesize;
          currentline=currentline+1;
          tmp=datevec (getdata (currentline ,1 ,A));
      end
      if vis_was_called
          pause (5);
        % F=getframe; %This line and next line meant for saving frames ...
                        % in file
        % imwrite (F.cdata, ['.\frames\frame_' regexprep(datestr (time), ':', ...
        % '-') '.png']);
```

```
     end % or pause (3) or some kind of check against current time
     time=time+1/24/60;
end

percent_bags_missed_flight=100*bags_missed_flights/bags_checked_in_for_flights
avg_time_to_storage=mean(times_from_check_in_to_storage*24*60)
avg_time_in_loop=mean(times_waiting_for_scan*24*60)
```

getdata.m

```
function s=getdata (line, col, file)
        x=5*(line-1)+ col;
        if (x>size (file, 1))
            s{1}= '\0\0';
        else
            s=file{x};
        end
```

handle_sensor_event.m

```
function handle_sensor_event (line, file)
global time
global tmp2
global weightthreshold
global bags_checked_in_for_flights
global bags_missed_flights
global times_from_checkin_to_storage
global times_waiting_for_scan
event_type=str2num(getdata (line,3, file));
switch event_type
%-------------------------------------------------------------------
  case 01 % photograph person who arrived with bag
    disp (['_Photo_taken_for_owner_of_bag_#' getdata (line,2, file)]);
    sensor_visualization (1, getdata (line,2, file),0, '\0\0', event type);
%-------------------------------------------------------------------
  case 02 % check if bag is over weight
    if str2num(getdata (line,4, file))> weightthreshold
        disp ([ 'Bag_#' getdata (line,2, file) '_overweight']);
        sensor_visualization (1, getdata (line,2, file), str2num(getdata ...
        (line, 4, file)), ' overweight ', event_type);
        % pass bag overweight event to visualization
    else
        disp ([ 'Bag_# ' getdata (line,2, file) '_weighed_in_at_' getdata ...
        (line, 4, file) ' lbs']);
        sensor_visualization (1, getdata (line,2, file), str2num(getdata ...
        (line, 4, file)), 'ok', event_type);
    end
%-------------------------------------------------------------------
  case 03
        disp ([ 'Bag_# ' getdata (line,2, file) '_scanned,_error=' getdata ...
        (line, 4, file )]);
```

```
        sensor_visualization (8, getdata (line,2, file),0, getdata ...
        (line,4, file), event_type);
        time_introduced_to_bhs=time;  %  random default; have not handled ...
                                      % case where bag was never checked in
        for x=1: line
              if strcmp (getdata (x,2, file), getdata (line,2, file)) ...
              && strcmp (getdata (x,3, file), '02')
                  tmp=datevec (getdata (x,1, file));
                  time_introduced_to_bhs=datenum ([tmp2 tmp (4:6)]);
              end
        end
        times_waiting_for_scan=...
        [times_waiting_for_scan time----time_introduced_to_bhs];
        % bags that were rescanned are currently averaged in twice.
        % possibly give 2nd scan different event id, otherwise need more logic
%-------------------------------------------------------------------
  case 04 % check if bag went in/out of storage
    results=0;
    for x=1:line
          if strcmp (getdata (x,2, file), getdata (line,2, file)) && strcmp (...
          getdata (x,3, file), '04')
              results=results+1;
          end
    end
sensor_visualization (10, getdata (line,2, file),0, '\0\0', event type);
if mod(results ,2)
    disp (['Bag_# ' getdata (line,2, file) '_went_into_temp_storage']);
    % into storage
    time_checked_in=time;  %  random default; have not handled ...
                           % case where bag was never checked in
    for x=1: line
          if strcmp (getdata (x,2, file), getdata (line,2, file)) && strcmp (...
          getdata (x,3, file), '01')
              tmp=datevec (getdata (x,1, file));
              time_checked_in=datenum ([tmp2 tmp (4:6)]);
          end
    end
    times_from_checkin_to_storage=...
    [times_from_checkin_to_storage time-time_checked_in];
else
    disp ([ 'Bag_# ' getdata (line,2, file) '_went_out_of_temp_storage']);
    % out of storage
end
%-------------------------------------------------------------------
  case 05
        disp ([ 'Bag_#' getdata (line,2, file) '_loaded_on_van_# ' getdata ...
        (line, 4, file)]);
        sensor_visualization (13, getdata (line,2, file),0, getdata ...
        (line,4, file), event_type);
%-------------------------------------------------------------------
  case 06 % check if bag that failed security scan was put on plane
      results=0;
      for x=1:line
```

```
        if strcmp (getdata (x,2, file), getdata (line,2, file)) && strcmp (...
        getdata (x,3, file), '03') && ...
                 strcmp (getdata (x,4, file), '0')
            results=results+1;
        end
    end
    if results>0
        disp ([ 'Bag_#' getdata (line,2, file) '_loaded_onto_flight_#' ...
        getdata (line,4, file)]);
        sensor_visualization (11, getdata (line,2, file),0, ' \0\0', ...
        event type);

    else
        disp ([ 'WARNING!' 10 'Bag_#' getdata (line,2, file) ...
         '_loaded_onto_flight_#' getdata (line,4, file) ...
                '_never_passed_security_inspection!']);
        sensor_visualization (11, getdata (line,2, file), str2num(getdata ...
        (line,4, file)), 'NOT_SCANNED', event_type);
    end
%-----------------------------------------------------------------------
  case 07 %flight left
    % query:#results ((field4==localdata(4)) && field3==01)-...
    % (field4==localdata(4)) && field3==06))>0
    bags_checked_for_current_flight=[ ];
    bags_loaded_on_current_flight=[];
    bags_lost=[];
    for x=1:line
      if strcmp (getdata (x,4, file), getdata (line,4, file)) && strcmp (...
      getdata (x,3, file), ' 01 ')
          bags_checked_for_current_flight=...
          [bags_checked_for_current_flight str2num(getdata (x,2, file))];
      end
      if strcmp (getdata (x,4, file), getdata (line,4, file)) && strcmp (...
      getdata (x,3, file), '06')
          bags_loaded_on_current_flight=[bags_loaded_on_current_flight...
           str2num(getdata (x,2, file))];
      end
    end
    bags_lost=intersect (setxor (bags_checked_for_current_flight, ...
    bags_loaded_on_current_flight), ...
    bags_checked_for_current_flight);
    bags_checked_in_for_flights=bags_checked_in_for_flights+size (...
    bags_checked_for_current_flight, 2);
    bags_missed_flights=bags_missed_flights+size(bags_lost, 2);
    if size (bags_lost,2)>0
        disp ([ 'WARNING! '10 'Bag(s) #' num2str (bags_lost)...
        '_not_loaded_onto_flight_#' getdata (line,4, file)]);
        for x=1:size (bags_lost,2)
            sensor_visualization (12,num2str (bags_lost (1, x)), 1, ...
            getdata (line,4, file), event_type);
        end
    end
    disp ([ 'Flight_#' getdata (line,4, file) '_departed.' ]);
    %write to global db ('Bag#', 'flight#', 'BagAttribute');
```

```
%-----------------------------------------------------------------------
  case 08 %flight arrived
      disp ([ 'Flight_#' getdata (line,4, file) '_arrived.']);
      %sensor_visualization (11, '\0\0',0, getdata (line,4, file), event_type);
      %download from global db ('Bag#', 'flight#', ' BagAttribute');
%-----------------------------------------------------------------------
  case 09 %Photograph taken
      disp ([ 'RFID_for_Bag_#' getdata (line,2, file) ...
      '_returned,_photo_captured ' ]);
      sensor_visualization (14,getdata (line,2, file),0, '\0\0 ', event type);
      % capture photo here
      % remove refrences to bag getdata (line,2, file) from local db,...
      % save in natl db
%-----------------------------------------------------------------------
  case 10 %Transfer red to Lost Baggage area
    results=0;
    for x=1:line
        if strcmp (getdata (x,2, file), getdata (line,2, file)) && strcmp (...
        getdata (x,3, file), '10')
            results=results+1;
        end
    end
sensor_visualization (15, getdata (line,2, file),0, '\0\0', event type);
if mod(results, 2)
    disp ([ 'Bag_#' getdata (line,2, file) '_went_into_lost_baggage_area']);
    % into lost baggage area
else
    disp ([ 'Bag_#' getdata (line,2, file) '_went_out_of_lost_baggage_area']);
    % out of lost baggage area
end
%-----------------------------------------------------------------------
  otherwise
    disp ([ 'ERROR_INVALID_EVENT_ON_LINE_' num2str(line)]);
end
```

sensor_visualization.m

```
function sensor_visualization (LOCATION,BAGID, number, special, event_type)
global DISPIMG
global viz_bag_loc_data;
global Environment
global time
Environment=imread ('maindiagram.jpg', 'jpg');
Environment=rgb2gray (Environment);

%Init coordinates
Desk1=[21 240]; % 1
Desk2= [21 338];% 2
Desk3= [21 440];% 3

Scanner1=[326 157]; % 4
Scanner2=[299 533]; % 5
```

```
Scanner3=[440 631]; % 6

XRAY1=[218 257];%7
XRAY2=[218 322];%8

LEVEL4INS=[173 659];%9

STORAGE=[480 590]; %10

PLANE=[540 100]; %11

LOSTAREA=[577 728]; %12

VANLOC=[550 300]; %13

PICKUPAREA=[150 690]; %14

LOSTBAGGAREA=[465 690]; %15


%Initializing coordinates
%LOCATION=4;
%BAGID=006;

if ~exist('DISPIMG ')
    DISPIMG=[];
    DISPIMG=Environment;
end

if ~exist ('viz_bag_loc_data')
    viz_bag_loc_data=[];
end

viz_bag_loc_data {size(viz_bag_loc_data,1)+1,1}=LOCATION;
viz_bag_loc_data {size(viz_bag_loc_data,1),2}=BAGID;
viz_bag_loc_data {size(viz_bag_loc_data,1),3}= number;
viz_bag_loc_data {size(viz_bag_loc_data,1),4}= special;
viz_bag_loc_data {size(viz_bag_loc_data,1),5}= event_type;


    switch LOCATION
    case 1
      DISPLOC=Desk1;
    case 2
      DISPLOC=Desk2;
    case 3
      DISPLOC=Desk3;
    case 4
      DISPLOC=Scanner1;
    case 5
      DISPLOC=Scanner2;
    case 6
      DISPLOC=Scanner3;
```

```
    case 7
      DISPLOC=XRAY1;
    case 8
      DISPLOC=XRAY2;
    case 9
      DISPLOC=LEVEL4INS;
    case 10
      DISPLOC=STORAGE;
    case 11
      DISPLOC=PLANE;
    case 12
       DISPLOC=LOSTAREA;
    case 13
        DISPLOC=VANLOC;
    case 14
        DISPLOC=PICKUPAREA;
    case 15
        DISPLOC=LOSTBAGGAREA;
    otherwise
        null;
    end

for i=(DISPLOC(1) -5):1:(DISPLOC(1)+5)
    for j=(DISPLOC(2) -5):1:(DISPLOC(2)+5)
           DISPIMG(i, j)=0;
    end
end

warning off;
imshow(DISPIMG);
%truesize;
warning on;
for i=1:1:size(viz_bag_loc_data, 1)
    LOCATION=viz_bag_loc_data{i,1};

    switch LOCATION
    case 1
      DISPLOC=Desk1;
    case 2
      DISPLOC=Desk2;
    case 3
      DISPLOC=Desk3;
    case 4
      DISPLOC=Scanner1;
    case 5
      DISPLOC=Scanner2;
    case 6
     DISPLOC=Scanner3;
    case 7
      DISPLOC=XRAY1;
    case 8
      DISPLOC=XRAY2;
```

```
case 9
  DISPLOC=LEVEL4INS;
case 10
  DISPLOC=STORAGE;
case 11
  DISPLOC=PLANE;
case 12
   DISPLOC=LOSTAREA;
case 13
    DISPLOC=VANLOC;
case 14
    DISPLOC=PICKUPAREA;
case 15
    DISPLOC=LOSTBAGGAREA;
otherwise
    null;
end
    offset=length (find (cell2mat (viz_bag_loc_data (1:i,1))==LOCATION))-1;
    text (DISPLOC(2)+10,DISPLOC(1)+6+18*offset, viz_bag_loc_data{i,2},...
    'fontsize' , 10);
    text (600,10,datestr(time), 'fontsize', 10);
    %-------------------------------
    if viz_bag_loc_data{i,5}==2
    text (DISPLOC(2)+50,DISPLOC(1)+6+18*offset,[num2str(...
    viz_bag_loc_data{i,3}) 'lbs.'], 'fontsize', 10);
        if  strcmp (viz_bag_loc_data{i,4}, 'overweight')
             text (DISPLOC(2)+100,DISPLOC(1)+6+18*offset, 'OVERWEIGHT',...
             'fontsize' ,10);
        end
    end
    %-------------------------------
    if viz_bag_loc_data{i,5}==3
        if  strcmp (viz_bag_loc_data{i,4}, '1')
             text (DISPLOC(2)+50,DISPLOC(1)+6+18*offset, 'FAILED',...
             'fontsize', 10);
        else
            text (DISPLOC(2)+50,DISPLOC(1)+6+18*offset, 'PASSED',...
            'fontsize', 10);
        end
    end
    %-------------------------------
     if viz_bag_loc_data{i,5}==6
        if  strcmp (viz_bag_loc_data{i,4}, 'NOT_SCANNED')
             text (DISPLOC(2)+50,DISPLOC(1)+6+18*offset, ['WARNING ! '...
              'Bag #' viz_bag_loc_data{i,2} '...
                               loaded onto flight #' ...
                               num2str (viz_bag_loc_data{i,3}) ' ...
                               never passed security inspection!'], 'fontsize', 10);
             for tmp=1:3
                 pause (.2);
                 beep;
             end
        else
```

```
                %text (DISPLOC(2)+50,DISPLOC(1)+6+18*offset, 'PASSED',...
                %'fontsize', 10);
            end
        end
        %-------------------------------
        if viz_bag_loc_data{i,5}==7
          text (DISPLOC(2)+10,DISPLOC(1)-20,['flight_#'...
          viz_bag_loc_data{i, 4}], 'fontsize', 10);
        end %--------------------------------
        if viz_bag_loc_data{i,5}==8
          %not visualizing right now
        end

end
```

```
sensor_visualization_clear.m
```

```
function sensor_visualization_clear ();
global DISPIMG
global viz_bag_loc_data;
global Environment

if ~size (DISPIMG, 1)
    Environment=imread ('maindiagram.jpg ',' jpg');
    Environment=rgb2gray (Environment);
end
DISPIMG=Environment;
viz_bag_loc_data=[];
```

PROBLEMS

15.1 Design an event generator that generates synthetic testing data for the BHS.

15.2 Propose a sensor network configuration applicable to geoscience problems that can also be extended to the social sciences or people-centric situations.

15.3 Describe an example integrating sensor networks and social networks.

15.4 Design a sensor-based tsunami warning system for tracking weather events in an unstructured enviornment.

15.5 Describe how to use sensor networks to implement monitoring of a costal erosion problem.

15.6 Apart from "Monitoring" activities, list three other important applications of sensor networks.

REFERENCES

1. Anonymous, Optimal data fusion in multiple sensor detection systems, *IEEE Trans. Aerospace Electron. Syst.* **AES-22**:98–101 (1988).
2. G. Black and V. Vyatkin, Intelligent component based automation of baggage handling systems with IEC 61499, *IEEE Trans. Autom. Sci. Eng.* **6** (2009).
3. V. T. Le, D. Creighton, and S. Nahavandi, Simulation-based input loading condition optimisation of airport baggage handling systems, *Proc. IEEE Intelligent Transportation Systems Conf.*, Seattle, WA, 2007.
4. K. Leone and R. Liu, The key design parameters of checked baggage security screening systems in airports, *J. Air Transport Mgmt.* **11**:69–78 (2005).
5. J. C. Rijsenbrij and J. A. Ottjes, New developments in airport baggage handling systems, *Transport. Plann. Technol.* **30**:417–430 (2007).

16 Security in Sensor Networks

If you spend more on coffee than on IT security, you will be hacked. What's more, you deserve to be hacked.

—White House Cybersecurity Advisor, Richard Clarke

16.1 INTRODUCTION

For the sake of completeness in the context of programming, this chapter provides a cursory view of security attacks and concerns in sensor network. For more details on algorithms and architecture, readers are advised to refer to some of the papers listed at the end of this chapter, as well as the rich literature listed in the Bibliography and available elsewhere.

16.2 SECURITY CONSTRAINTS

The fundamental constraints under which sensor networks operate prohibits them from using public-key encryption systems and third-party authentication systems. These constraints are described in the following subsections.

16.2.1 Resource Constraints

Resource constraints drive every aspect of sensor programming. As noted, the low power and processing capabilities of sensors are the most significant factors in sensor security. A typical sensor node might have a maximum of around 20–30 J (joules) of energy. For example, a Berkeley mote has an 8-bit, 4-MHz processor, which supports a minimal reduced instruction set computer (RISC)-like instruction set without support for multiplication or other costly operations. Perrig et al. [1, 2] showed that a simple random structures–algorithms (RSA) operation takes on the order oftens of seconds on this processor. After a mote is loaded with the requisite os and communications software, it has less than 4 kB of free space. With space at such a high premium, it is not possible to store too many long keys, or even long algorithms.

16.2.2 Communication Issues

Standard sensor nodes equipped with low-power radio transmitters commonly have a range of under 20 m. In order for these nodes to communicate with a base station any significant distance away, they need to use multihop routing. Because the base station doesn't communicate directly with most of the nodes in its WSN, it cannot efficiently manage key distribution for its nodes because of the high overhead that this would require. Also, security in traditional networks depends on well-established protocols that assume that a reliable media exists for any kind of authentication to take place. In sensor networks, issues of latency, conflicts, and unreliability of the underlying media arise, rendering most authentication schemes suboptimal for these classes of network.

16.2.3 Hostile Environments

Distributed sensor networks (DSNs) that are deployed in hostile environments must account for the fact that individual nodes can be easily captured. A captured public-key server could potentially be used to disclose a large number of keys, compromising the network. Finally, the intrinsic properties of sensor networks (distributed in nature and avoiding central management) makes it challenging to adopt authentication schemes developed for more conventional networks.

16.2.4 Conclusion

For a broader treatment of these subjects, refer to the article by Kalindi et al. [3].

16.3 DENIAL-OF-SERVICE ATTACKS IN MULTIPLE LAYERS

Sensor networks could be subjected to several kinds of attacks not limited to denial-of-service attacks, node takeovers, routing attacks, and possibly attacks on the physical security of nodes. This section provides an overview of potential denial-of-service attacks that can take place at different layers of the sensor network. It is important to note that in each layer of the sensor network architecture, the denial-of-service attack is of a different nature (see Fig. 16.1).

16.3.1 Physical Layer

Two types of denial-of-service attacks take place at the physical layer: jamming and tampering.

Jamming *Jamming* refers to the transmission of radio signals that interfere with the communication operations of the nodes in a sensor network. Jamming is a simple and effective method of disrupting communications. Generating a significant amount of noise on a narrow band of frequences will disrupt the communications that use those

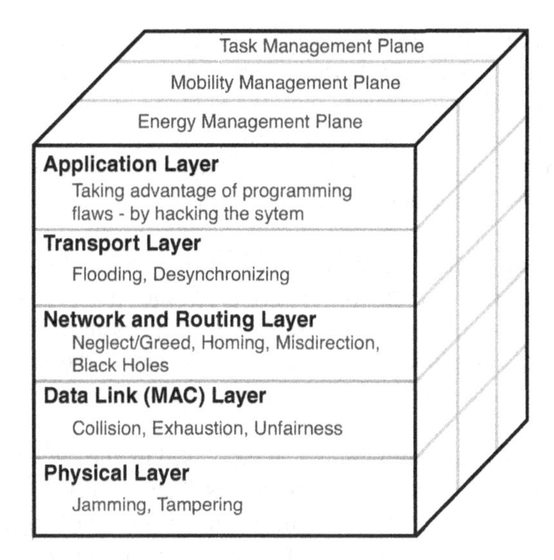

FIGURE 16.1 An illustration showing possible attack types on each layer.

frequences. Jamming can be eliminated by changing frequencies, or by using a wide band of frequencies via modern spread-spectrum techniques. Figure 16.2 illustrates how a jammer could operate in a wireless sensor network.

Tampering *Tampering* is when the network hardware is physically attacked, damaged, or otherwise compromised. The best defense against tampering is hiding the nodes, or making them physically resistant to attack.

16.3.2 Datalink Layer

A collision occurs when two packets are sent on the same channel at the same time, corrupting the recieved data. Because radio is inherently a half-duplex system, a node cannot easily receive during transmission in order to check for collisions. Furthermore, even if this were possible, the signal received at a node's neighbors would be significantly different from that at point-blank range. So in wireless communications,

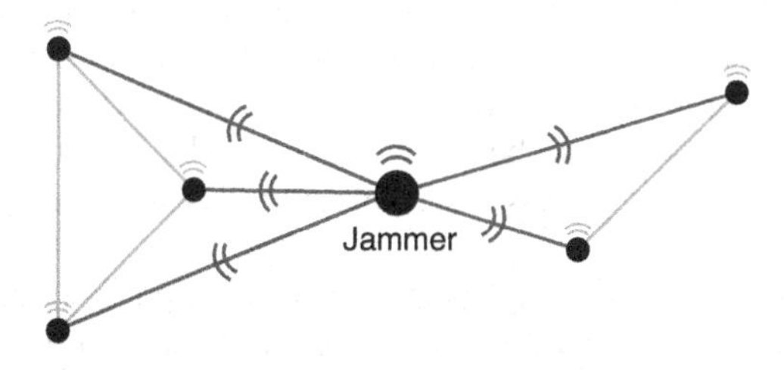

FIGURE 16.2 An illustration of a jammer in a sensor network.

a collision appears for all intents and purposes as corrupted data at the reciever's end. Unfortunately, if the attacker has already compromised one or more nodes and is able to send malicious packets onto the network at will, there is no fully effective solution. If the collisions are corrupting only a portion of the sent data packets, this issue can be mitigated by using error-correcting codes. Some form of collision detection at the receiver could be used to detect such attacks, if not to completely mitigate them.

An exhaustion attack is similar to a collision attack, but the aim of the attack is to use up the power of network nodes and render them inoperative. One way to accomplish this is to purposely cause repeated collisions and retransmissions. Another way is to have a compromised node continually query other nodes for information. Limiting the maximum amount of reponses that a node is allowed to make in a given time period can mitigate this problem.

Network delays and unreliability can be caused by the intermittent application of the abovementioned attacks. In an "unfairness" attack, malicious users transmit an unusually large number of packets if the medium is free, causing other nodes to delay sending their packets. This type of attack relies on abusing the cooperative priority scheme used by the MAC layer. One way to mitigate this type of attack is to use small frames, causing nodes to give up control of the channel after only a short time. This does have overhead, however, and furthermore, a node that cheats by vying for the channel immediately while other nodes delay randomly can still capture the channel repeatedly.

16.3.3 Network Layer

A neglectful node will sometimes arbitrarily fail to forward data to the next node along the routing path. A compromised node could even falsely acknowledge successful transmission to the sender of the dropped packet. A "greedy" node gives higher priority to its own messages, causing a hit to network performance if a high volume of network traffic passes through it. One way around this is to use data redundancy, or better, muliple routing paths.

A "homing" attack is an attack where adversaries attempt to discover and focus on nodes that are particularly crucial or important, such as local clusterheads and, key managers. This type of attack can be foiled if packet headers are encrypted, preventing adversaries from easily finding the source or destination of a given packet.

A "misdirection" attack involves routing data along the wrong path, possibly by advertising false routing information. As a result, data could be routed away from a specific node, or sent toward the malicious node instead of their intended destination. In a "smurf" attack, a large amount of data is mistakenly routed toward the node being attacked, overloading its network connection. An approach similar to egress filtering can be used to defend against this type of attack.

A "black hole" is a node that falsely advertises zero-cost routes to all other nodes, causing all nearby nodes to forward packets to it, and to incorrectly update their own route costs to be erroneously low, causing nodes in a continually expanding

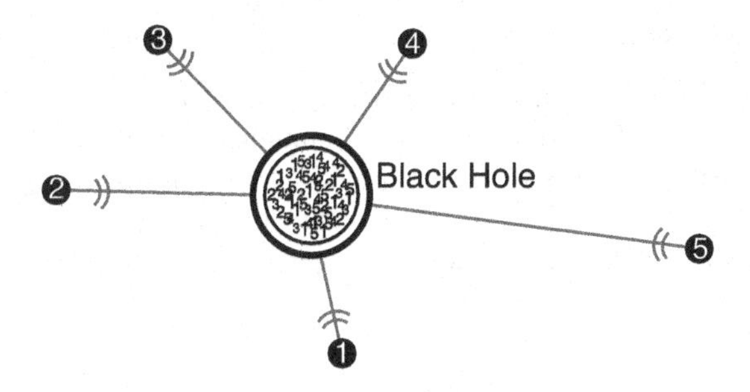

FIGURE 16.3 Packets being routedtoa black hole.

radius to route their data toward the black hole (see Fig.16.3). This attack can be detected by looking for suspicious routing cost claims, and if detected, is fairly easy to defend against. However, if not detected, it can be extremely disruptive. Having nodes monitor their neighbors provides a good defense against this type of attack, although with overhead. Occasional network probing can also be used to locate blackout areas, and distributed probing schemes are possible. It is important that a probe be indistinguishable from ordinary traffic, or a malicious node could appear benign during probing.

"Flooding" is when many packets are sent to a node in an attempt to overload it. One example of this is the SYN flood, where an attacker sends a large number of requests for connections, binding up all the target's resources on pending transactions. Similar types of attack are possible on WSNs, where an adversary can waste a node's resources by sending many connection requests. Limiting the number of incoming connections will save resources, but service of legitimate connection requests will still be slow at best. A better option is to make it computationally costly for a node to make a connection establishment request. "Client puzzles" are simple mathematical puzzles that a client needs to solve before asking to establish a connection. In this way, a node is prevented from generating many extraneous requests at once.

16.3.4 Transport Layer

A desynchronization attack occurs when a malicious attacker uses falsified messages to cause two nodes to believe that they are out of sync, and repeatedly run a synchronization recovery protocol. If the packets being received at the endpoints can be authenticated, this attack will not work.

16.3.5 Application Layer

Attacks at this layer attempt to disrupt or crash individual services running on the target computer, denying service in that way. Many traditional hacking methods rely on exploiting software vulner abilities in this manner.

16.3.6 Conclusion

Many of the aformentioned attacks could be prevented by using encryption to authenticate data packets. Unfortunately, the limitations of small sensor nodes, and those of ad hoc sensor networks in general, make such approaches largely unfeasible. As in most programming problems, security threats to sensor networks can be handled most efficiently if they are considered during design, so security-aware design is very important.

16.4 SOME WELL-KNOWN ALGORITHMS FOR SECURITY PROBLEMS

The following paragraph describes an overview of some of the known algorithms in the security aspects of sensor networks. For the purpose of generality, we are providing only an overview of these security algorithms. For more details and information on related projects, refer to the references listed at the end of this chapter.

16.4.1 KKID: Sub-Grid-Based Key Vector Assignment: A Key Predistribution Scheme for Sensor Networks

In general, a secure sensor network framework is of utmost importance as these vary sensors are placed in environments that pose a high risk of sensor capture and perhaps, destruction. In addressing this issue, the employment of certain preventive mechanisms such as trusted third-party authentication and public-key systems render useless as these mechanisms oftentimes exhibit sub-optimal resource requirements. Key predistribution was introduced in 2003 [2] to solve this problem. Our scheme achieves connectivity identical to that of random key predistribution [2] but fewer using preloaded keys in each sensor node. The design of our scheme is motivated by the observation that at present most key predistribution schemes employ random mechanisms that use a large number of keys and are unsuitable for sensor networks. In this algorithm we extend the deterministic key predistribution scheme that we proposed in our earlierwork [3], which is based on assigning keys to sensors by placing them on a grid. This approach has been further modified to use multiple mappings of keys to nodes. In each mapping every node receives a distinct set of keys that it shares with different nodes. The key assignment is done such that there will be keys in common between nodes in different subgrids. After being randomly deployed, the nodes discover common keys, authenticate, and communicate securely. The analysis and simulation results show that this scheme is able to achieve better security compared to the random schemes.

For a full treatment on this topic, refer to the article by Kalindi et al. [3].

16.5 SECURE INFORMATION ROUTING

Nodes deployed in hostile environments are prone to capture. Capture of a single node discloses all the information about the keys that they contain. More specifically, an adversary can capture multiple nodes by eavesdropping on radio transmissions, injecting bits into the channel, and repeating previously heard packets. Adversary nodes can be about as powerful as existing nodes, or significantly more powerful. The paper by Karlof and Wagner [4] presents a threat model based on two classes of attacker: mote class attacker, where the attacker has access to a few sensor nodes similar to legitimate nodes, and laptop class attacker, where adversaries have access to more powerful computational power, more battery power, and a high-powered radio transmitter. This algorithm presents a secure routing protcol that guarantees protection against eavesdropping, integrity, authenticity, and availability of messages. For more details, refer to Ref. [4].

16.6 SECURITY PROTOCOLS FOR SENSOR NETWORKS

This particular algorithm has many features, such as data confidentiality, which includes encryption data with shared key, data authentication, and integrity. This scheme allows the receiver to verify whether the data were sent by the client sender, and also guarantees that messages are not altered in transit by hostile attackers. One unique point of this algorithm is data freshness, which guarantees that no adversaries replayed old sensor readings. It also has a certain amount of odering of the sensor data due to data esimation. A good mathematical theory has been developed for this sensor network encryption protocol for authenticated broadcast of network data. For a broader treatment on this algorithm, the reader can refer to the article by Perrig et al. [5].

16.7 FINAL COMMENTS

Security in sensor networks is very critical to enhancing the long-term usefulness of sensor networks for various applications. This chapter has given a brief overview of the security aspects of these networks. By no means is this a complete treatment of the subject matter. Furthermore, sensor networks is an area of national importance for many defense and civilian applications, and cannot be considered deployable without sufficient protection from denial-of-service and other major attacks. Consideration of sensor network security in the design phase of the network can certainly ensure successful network deployment down the line, and head off problems before they occur.

PROBLEMS

16.1 Develop a protocol by formulating a deterministic key predistribution scheme proposed by the KKID algorithm [3] that is based on assigning keys to sensors by placing them on a grid.

16.2 Develop a programming tool by using a cryptographic authentication mechanism, by attempting to add denial-of-service resistance to existing protocols.

16.3 What are the programming constraints in developing a secure sensor network to be deployed in a hostile environment that is prone to malicous attacks?

16.4 Why is effective collision detection problematic in wireless networks?

16.5 Give an example of a problem/algorithm that could be used as a client puzzle, and implement it in a program.

REFERENCES

1. J. Stankovic, A. Perrig, and D. Wagner, Security in wireless sensor networks, *Commun. ACM* **47**:53–57 (2004).
2. H. Chan, A. Perrig, and D. Song, Random key predistribution schemes for sensor networks, *Proc. IEEE Security and Privacy Symp. 2003* (May 2003).
3. R. Kalindi, R. Kannan, S. S. Iyengar, and A. Durresi, Sub-grid based key vector assignment: A key pre-distribution scheme for distributed sensor networks, *J. Pervasive Comput. Commun.* **2**(1):35–43 (March 2006).
4. C. Karlof and D. Wagner, Secure routing in wireless sensor networks: Attacks and countermeasures, *Proc. 1st Int. IEEE Workshop on Sensor Network Protocols and Applications*, Univ. California, Berkeley, 2003.
5. A. Perrig, R. Szewczyk, V. Wen, D. Culler, and J. D. Tygar, Spins: Security protocols for sensor networks, *Wireless Networks* **8**(5):521–534 (2002).

17 Closing Comments

A successful application of sensor networks is to prevent possible future disasters by analyzing the sensor-based data collected over long periods of time. A case in point is tsunami warnings. Tsunamis can be described as very long-wavelength waves of water caused by a sudden displacement of the ocean bed. The rate at which a wave loses energy is inversely related to its wavelength and its velocity, and directly proportional to water depth. Most tsunamis are caused by undersea earthquakes, volcanic eruptions, landslides, or meteor impacts, and are usually preceded by seismic disturbances. In case of inland seismic events, there exists a worldwide network of sensors. A similar network is conspicuously absent for events originating at sea. Sensor-based tsunami warning systems have proved to be effective in Japan and the United States. The currently operational system for tsunami detection, called "deep-ocean assessment and reporting of tsunami" (DART) is very useful for the tracking of tsunami warnings. The distributed sensor network will certainly advance the state of the art in wireless tsunami-based sensor networks by designing in expensive, expendable, and massively deployable innovative sensors.

It would be very interesting to integrate these types of sensor networks with social networks where cellular sensors can infiltrate people's everyday lives, thereby providing real-time information about their surroundings. Thus, online sensor-based networks will have tremendous future in many of these applications.

Bibliography

Agre, J., L. Clare, and S. Sastry, A taxonomy for distributed real-time control systems, *Adv. Computer.* **49:**303–352 (1999).

Akyildiz, I. F., W. Su, E. Cayirici, and Y. Sankarasubramaniam, A survey of sensor networks, *IEEE Commun. Mag.*, **8:**102–114 (2002).

Akyildiz, I. F., W. Su , Y. Sankarasubramaniam, and E. Cayirci, Wireless sensor networks: A survey, *Comput. Networks* **38:**393–422 (2002).

P802.11k (C/LM) Amendment to STANDARD [FOR]. *Information Technology—Telecommunication and Information Exchange between Systems—Local and Metropolitan Area Networks Specific Requirements—Part 11: Wireless LAN medium access control (MAC) and Physical Layer (PHY) Specifications,* Radio Resource Measurement of Wireless LANs, 1999.

Anonymous, Optimal data fusion in multiple sensor detection systems, *IEEE Trans. Aerospace Electro. Syst.*, **AES-22:**98–101 (1988).

Basavaraju, S., *Sensim: A Wireless Sensor Network Simulation Template,* M.S. Project, Dept. Computer Science, Louisiana State Univ. Baton Rouge.

Bharghavan, V., A. Demers, S. Shenker, and L. Zhang, Macaw: A media access protocol for wireless LANS, *Proc. ACM SIGCOMM 1994,* 1994.

Black, G. and V. Vyatkin, Intelligent component based automation of baggage handling systems with IEC 61499, *IEEE Trans. Autom. Sci. Eng.* **6**(2009).

Bondy, J. A. and U. S. R. Murty, *Graph Theory with Applications*, North Holland, NewYork, 1976.

Brooks, R. R. and S. S. Iyengar, *Multi-Sensor Fusion*, Prentice-Hall, Englewood Cliff, NJ, 1997.

Cassandras, C. G. and S. Lafortune, *Introduction to Discrete Event Systems,* Kluwer Academic, Jan. 1999.

Chakrabarty, K. and S. S. Iyengar, *Scalable Infrastructure for Distributed Sensor Networks*, Springer-Verlag, 2005.

Chan, H., A. Perrig, and D. Song, Random key predistribution schemes for sensor networks, *Proc. IEEE Security and Privacy Symp. 2003,* May 2003.

Chandrasekharan, N. and S. Iyengar, NC algorithms for recognizing chordal graphs and k-trees, *IEEE Trans. Comput.* **37:**10 (1988).

Crossbow Imote2 Datasheet, courtesy Crossbow Technologies.

Crossbow MIB520 Datasheet, courtesy Crossbow Technologies.

Crossbow Moteworks Software Reference Manual, courtesy Crossbow Technologies.

Crossbow Product Feature Reference Manual, courtesy Crossbow Technologies.

Crossbow Reference Manual, courtesy Crossbow Technologies.

Crossbow Telosb Datasheet, courtesy Crossbow Technologies.

Eckmann, S. T., G. Vigna, and R. A. Kemmerer, Statl: An attack language for state-based intrusion detection, *Proc. ACM Workshop on Intrusion Detection*, Nov. 2000.

Eschenauer, L. and V. D. Gligor, A key management scheme for distributed sensor networks, *Proc. 9th ACM Conf. Computer and Communication Security*, Nov. 2002, pp. 41–47.

Eskin, E. and W. Lee, Modeling system calls for intrusion detection with dynamic window sizes, *Proc. DISCEX II*, 2001.

Fall, K. and Varadhan, *Ns-2 Network Simulator,* Technical Report, Univ. California, Berkeley, 2004.

Fishman, G. S., *Principles of Discrete Event Simulation*, Wiley, 1978.

Forrest, S., C. Warrender, and B. Pearlmutter, Detecting intrusions using system calls: Alternative data models, *Proc. 1999 IEEE Symp. Security and Privacy,* IEEE Computer Society, 1999, pp. 133–145.

Gislason, D., *ZigBee Resource Guide*, Webcom Communication Corpo., 2008.

Global Sensor Networks, GSNTeam. http://sourceforge.net/projects/gsn/

Hill, J., R. Szewczyk, A. Woo, S. Hollar, D. Culler, and K. Pister, System architecture directions for networked sensors. *ACM Sigplan Notices,* **35:**93–104 (2000).

Hill, J., R. Szewczyk, A. Woo, S. Hollar, D. Culler, and K. Pister, System architecture directions for networked sensors. *In Architectural Support for Programming Languages and Operating Systems*, 2000, pp. 93–104.

Hofmeyr, S. A., S. Forrest, and A. Somayaji, Intrusion detection using sequences of system calls, *J. Comput. Security*, **6**(3):151–180 (1988).

Hopcraft, J. E., R. Motwani, and J. D. Ullman, *Introduction to Automata Theory, Languages, and Computation,* 2nd ed., Addison-Wesley Nov. 2001.

http://blog.xbow.com/xblog/sensorboards.

http://inst.eecs.berkeley.edu/cs194-5/sp08/lab1/index.html.

http://www.cs.rpi.edu/cheng3/sense/.

http://www.isi.edu/nsnam/ns/ns-documentation.html.

IEEE Standard Dictionary of Electrical and ElectronicTerms, 6th ed., IEEE, 1997.

Ilgun, K., R. A. Kemmerer, and P. A. Porras, State transition analysis: A rule-based intrusion detection approach, *IEEE Trans. Software Eng.* **21**(3):151–180 (March 1995).

Intanagonwiwat, C., R. Govindan, D. Estrin, J. Heidemann, and F. Silva, Directed diffusion for wireless sensor networking, *IEEE/ACM Trans. Networking* **11**(1):216 (Feb. 2003).

Iyengar, S. S. and R. R. Brooks, eds., *Distributed Sensor Networks,* 2nd ed., CRC Press, Dec. 2004.

Iyengar, S. S. and R. Brooks, eds., *Distributed Sensor Networks*, CRC Press, 1995.

Iyengar, S. S., R. L. Kayshyap, and R. N. Madan, Distributed sensor networks, *IEEE Trans. Syst. Man Cyber.* **21**(5):1027–1031 (1991).

Iyengar, S. S., L. Prasad, and H. Min, *Advances in Distributed Sensor Integration: Application and Theory*, Prentice-Hall, 1995.

Iyer, V., R. M. Garimella, Rama Murthy, and M. B. Srinivas, Min loading max reusability fusion classifiers for sensor data model, *Proc. 2nd Int. Conf. Sensor Technologies and Applications, SENSORCOMM '08,* Aug. 25–31, 2008, pp. 480–485.

Iyer, V., S. S. Iyengar, N. Balakrishnan, V. Phoha, and M. B. Srinivas, FARMs: Fusionable ambient renewable MACs. *Proc. IEEE Sensors Applications Symp. SAS 2009,* Feb. 17–19, 2009, pp. 169–174.

Iyer, V., S. S. Iyengar, G. Rama Murthy, M. B. Srinivas, and B. Hochet, Multi-hop scheduling and local datalink aggregation dependent qos in modeling and simulation of power-aware wireless sensor networks, Proceedings of 2009. ACM-IWCMC, Leipzig, Germany, 2009; pp. 844–848.

Iyer, V., G. Rama Murthy, M. B. Srinivas, and B. Hochet, C-error simulator for development for sensor and location-aware sensing applications, *Proc. 3rd Int. Conf. Sensing Technology, ICST 2008, Nov. 30–Dec. 3,* 2008, pp. 192–199.

Iyer, V., R. Murthy, M. B. Srivinas, and B. Hochet, Training data compression algorithms and reliability in large wireless sensor networks, *SUTC Proc. IEEE Int. Conf. Sensor Networks, Ubiquitous and Trustworthy Computing,* June 2008, pp. 480–485.

Iyer, V., G. Rama Murthy, and M. B. Srinivas, Training data compression algorithms and reliability in large wireless sensor networks, *Proc. IEEE Int. Conf. Sensor Networks, Ubiquitous and Trustworthy Computing,* June 2008, pp. 480–485.

Iyer, V., G. Rama Murthy, and M. B. Srinivas, Environmental measurement OS for a tiny CRF-stack used in wireless network, *Modern Sensing Technol.* (special issue) **90**:72–86 (2008).

Johnson, D. B., *The Rice University Monarch Project,* Technical Report, Rice Univ., 2004.

Johnson, D. S., The NP-completeness column: An outgoing guide, *J. Algorithms,* **6:**434–451 (1985).

Kalidindi, R., V. Parachuri, S. Basavaraju, C. Mallanda, A. Kulshrestha, L. Ray, R. Kannan, and A. Durresi, Sub-grid based key vector assignment: A key pre-distribution scheme for distributed sensor networks, *ICWN,* 2004.

Kalidindi, R., R. Kannan, S. S. Iyengar, and A. Durresi, Sub-grid based key vector assignment: A key pre-distribution scheme for distributed sensor networks, *J. Pervasive Comput. Communi.* **2**(1):35–43 (March 2006).

Karl, H. and A. Willig, Protocols and Architectures for Wireless Sensor Networks, John Wiley & Sons, Inc., 2005.

Karlof, C. and D. Wagner, Secure routing in wireless sensor networks: Attacks and countermeasures, *Proc. 1st Int. IEEE Workshop on Sensor Network Protocols and Applications, Univ. California, Berkeley,* 2003.

Karsai, G., A. Ledeczi, J. Sztipanovits, G. Peceli, G. Simon, and T. Kovacshazy, An approach to self adaptive software based on supervisory control, *Proc. Int. Workshop on Self Adaptive Software,* 2001.

Klein, P. N., *Efficient Parallel Algorithms for Planar, Chordal and Interval Graphs*, Ph.D. thesis, MIT, Cambridge, MA, 1988.

Klein, P. N., Efficient parallel algorithms for chordal graphs, *Proc. IEEE 29th Symp. Foundation of Computer Section* 1988, pp. 150–161.

Kumar, R. and M. Fabian, Supervisory control of partial specification arising in protocol conversion, *Proc. 35th Allerton Conf. Communication, Control and Computing,* 1997, pp. 543–552.

Kumar, R. and V. Garg, *Modeling and Control Logical Discrete Event Systems*, Kluwer Academic, 1995.

Kumar, S. and E. H. Spafford, A generic virus scanner in C++, *Proc. 8th Computer Security Applications Conf.,* 1992.

Le, V. T., D. Creighton, and S. Nahavandi, Simulation-based input loading condition optimisation of airport baggage handling systems, *Proc. IEEE Intelligent Transportation Systems Conf.,* Seattle, WA 2007.

LeCharlier, B. and M. Swimmer, Dynamic detection and classification of computer viruses using general behavior patterns, *Proceedings of 5th Int. Virus Bulletin Conf.* Sept. 1995, p. 75.

Lee, W. and S. J. Stolfo, Data mining approaches for intrusion detection, *Proc. 7th USENIX Security Symp. SECURITY '98*, Jan. 1998.

Leone, K. and R. Liu, The key design parameters of checked baggage security screening systems in airports, *J. Air Transport Mgmt.* **11**:69–78 (2005).

Levin, R. B., *The Computer Virus Handbook*, Osborne/McGraw-Hill, 1990.

Levis, P. and D. Gay, *TinyOs Programming*, Cambridge Univ. Press, 2009.

Linz, P., *An Introduction to Formal Languages and Automata,* 3rd ed., Jones & Barlett, Oct. 2000.

LSU Research Group, *LSU Sensor Simulator* (LSU SenSim, version 1, Jan. 2005) *User Manual*, Dept. Computer Science, Louisiana State University, Baton Rouge.

Mallanda, C., *Sensor Simulator: A Simulation Framework for Sensor Networks,* master's thesis, Dept. of Computer Science, Louisiana State Univ., Baton Rouge.

Michael, C. and A. Ghosh, Using finite automata to mine execution data for intrusion detection: A preliminary report, *Lect Notes Comput Sci,* **1907/2000**:66–79(2000).

Misra, J., Distributed discrete-event simulation, *ACM Comput. Surveys* **18**(1):39–65 (March 1986).

Moitra, A. and S. S. Iyengar, Parallel algorithms for some computational problems, *Adv. Comput.* **26**:93–153 (1987).

Nuansri, N., S. Singh, and T. S. Dillon, A process state-transition analysis and its application to intrusion detection, *Proc. ACSAC1999,* 1999, pp. 378–388.

OPNET Technolgies, Inc., *Opnet Modeler*. www.opnet.com

Park, S., A. Savvides, and M. B. Srivastava, Sensorsim: A simulation framework for sensor networks, *Proc. 3rd ACM Int. DRAFT Workshop on Modeling, Analysis and Simulation of Wireless and Mobile Systems,* 2000, pp. 104–111.

Perrig, A., R. Szewczyk, V. Wen, D. Culler, and J. D. Tygar, Spins: Security protocols for sensor networks, *Wireless Networks* **8**(5):521–534 (2002).

Polastre, J., J. Hill, and D. Culler, Versatile low power media access for wireless sensor networks, *Proc. 2nd Int. Conf. Embedded Networked Sensor Systems, SenSys '04,* ACM, New York, 2004, pp. 95–107.

Ramadge, P. J. and W. M. Wonham, Supervisory control of a class of discrete event processes, *SIAM J. Control and Optim.,* **25**(3):206–230 (1987).

Research Integration: Platform Survey, embedded WiSeNts consortium.

Rhee I., A. Warrier, M. Aia, J. Min, and M. L. Sichitiu, *Z-mac: A Hybrid MAC for Wireless Sensor Networks,* 2008, IEEE Press, Piscataway, NJ, vol. **16**, pp. 511–524.

Rijsenbrij, J. C. and J. A. Ottjes, New developments in airport baggage handling systems, *Transport. Plann. Technol.* **30**:417–430 (2007).

Ruiz-Sandoval, M., T. Nagayama, and B. F. Spencer, Sensor development using Berkeley mote platform, *J. Earthquake Eng.,* **10**:289–309 (2006).

Sastry, S., Smart space for automation, *Assembly Autom.,* **24**(2):201–209 (2004).

Sastry, S., S. S. Iyengarand, and N. Balakrishnan, Sensor technologies for future automation systems, *Sensor Lett.* **2**(1):9–17 (2004).

Sastry, S. and S. S. Iyengar, *Distributed Sensor Networks*, CRC Press, 2005.

Sastry, S. and S. S. Iyengar, *A Taxonomy of Distributed Sensor Networks*, CRC Press, 1995.

Schultz, M. G. and E. Eskin, et al., Data mining methods for detection of new malicious executables, *Proc. IEEE Symp. Security and Privacy, May* 2001.

Sobieh, A. and J. C. Hou, *A Simulation Framework for Sensor Networks in j-sim,* Technical Report UIUCDCS-R2003-2386, Dept. Computer Science, Univ. Illinois, Urbana–Champaign, Nov. 2003.

Solomon, A. and T. Kay, *Dr. Solomon's PC Anti-virus Book*, Newtech, 1994.

Spinellis, D., Trace: A tool for logging operating system call transaction, *Operating Syst. Rev.* **28**(4):56–63 (Oct. 1994).

Srinivas, M. B., V. Iyer, G. Rama Murthy, and B. Hochet, C-error simulator for development for sensor and location aware sensing applications, *Proc. 3rd Int. Conf. Sensing Technology,* Taichung, Taiwan, 2002, pp. 799–804.

Stankovic J., A. Perrig, and D. Wagner, Security in wireless sensor networks, *Commun. ACM* **47**:53–57 (2004).

Tannenbaum, A. S., *Computer Networks*, Prentice-Hall, 2002.

Vargas, A., *Omnet++ Discrete Event Simulation System,* version 2.3, 2003.

Vieira, M. A. M., D. C. da Silva Jr., C. N. Coelho Jr., and J. M. da Mata, Survey on wireless sensor network devices, *Proc. IEEE Conf. Emerging Technologies and Factory Automation, ETFA03,* 2003, p. 1.

Wallace, C., P. Jensen, and N. Soparkar, Supervisory control of workflow scheduling, *Proc. Int. Workshop on Advanced Transaction Models and Architectures,* 1996.

Wood, A. and J. A. Stankovic, Denial of service in sensor networks, *IEEE Comput.* **35**(10):54–62 (Oct. 2002).

Xavier, C. and S. S. Iyengar, *Introduction to Parallel Algorithms*, Wiley, 1998.

Ye, W., F. Silva, and J. Heidemann, Ultra-low duty cycle MAC with scheduled channel polling, *SenSys '06, Proc. 4th Int. Conf. Embedded Networked Sensor Systems,* ACM, NewYork, 2006, pp. 321–334.

Yu, Y., R. Govindan, and D. Estrin, *Geographical and Energy Aware Routing: A Recursive Data Dissemination Protocol for Wireless Sensor Networks,* Technical Report, Aug. 2001.

Zeng, X., R. Bagrodia, and M. Gerla, Glomosim: A library for parallel simulation of large-scale wireless networks, *Proc. Workshop on Parallel and Distributed Simulation,* 1998, pp. 154–161.

ZigBee Wireless Networking, Newnes Publications, 2008.

Index

Printed in the United States
By Bookmasters